Cybercrime Risks and Responses

Palgrave Macmillan's Studies in Cybercrime and Cybersecurity

This book series addresses the urgent need to advance knowledge in the fields of cybercrime and cybersecurity. Because the exponential expansion of computer technologies and use of the Internet have greatly increased the access by criminals to people, institutions, and businesses around the globe, the series will be international in scope. It provides a home for cutting-edge long-form research. Further, the series seeks to spur conversation about how traditional criminological theories apply to the online environment. The series welcomes contributions from early career researchers as well as established scholars on a range of topics in the cybercrime and cybersecurity fields.

Series Editors:

MARIE-HELEN MARAS is Associate Professor and Deputy Chair for Security at the Department of Security, Fire, and Emergency Management at John Jay College of Criminal Justice, USA.

THOMAS J. HOLT is Associate Professor in the School of Criminal Justice at Michigan State University, USA.

Titles include:

Cybercrime Risks and Responses

Eastern and Western Perspectives

Edited by

Russell G. Smith
Australian Institute of Criminology

Ray Chak-Chung Cheung
City University of Hong Kong, Hong Kong SAR

Laurie Yiu-Chung Lau
Asia Pacific Association of Technology and Society, Hong Kong SAR

First published 2015 by
PALGRAVE MACMILLAN

Palgrave Macmillan in the UK is an imprint of Macmillan Publishers Limited,
registered in England, company number 785998, of Houndmills, Basingstoke,
Hampshire, RG21 6XS.

Palgrave Macmillan in the US is a division of St Martin's Press LLC,
175 Fifth Avenue, New York, NY 10010.

Palgrave Macmillan is the global academic imprint of the above companies and
has companies and representatives throughout the world.

Palgrave® and Macmillan® are registered trademarks in the United States,
the United Kingdom, Europe and other countries.

ISBN: 978-1-137-47415-5

This book is printed on paper suitable for recycling and made from fully
managed and sustained forest sources. Logging, pulping and manufacturing
processes are expected to conform to the environmental regulations of the
country of origin.

A catalogue record for this book is available from the British Library.

A catalog record for this book is available from the Library of Congress.

For
Melinda,
Vicky, and Raisie
and
Lau Cheung Kwai Lan and Lam Ding Kwong, at peace

Contents

List of Figures

List of Tables

Foreword

In the summer of 2013, the International Conference on Cyber Crime and Computer Forensic was held by the Asia Pacific Association of Technology and Society (APATAS) in Hong Kong, and I had the privilege of meeting experts and researchers in the field of cybercrime. Several important themes have emerged from that dialogue. The proliferation of cybercrime has grown at an unprecedented speed, and as a Legislative Council member representing the information technology sector and an advocate for a free and open Internet, I am glad to witness the immense effort put into the preparation of this volume of research addressing Eastern and Western perspectives on cybercrimes.

Cybercrimes are increasingly sophisticated and organized, and the stakes are becoming higher. As shown in the hacks of bitcoin exchange and a US film company, skilled terrorist groups and hacktivists are becoming more active. Cybercrime has also evolved into a tool of chasing big money, stealing confidential information, and even making a political stand. The days when the stereotypical hackers were curious, tech-savvy coders trying to prove their skills are now far behind. Cybercrime has evolved in scale and format, with both unauthorized access to information and dissemination of illegal content being common across borders. The cost of cybercrime to individuals, corporations, and governments skyrockets, and the resources needed to trace, attribute, and convict criminals have also grown. Meanwhile, authorities and law enforcement agencies are often outpaced by the speed and skill of criminals, making policing an outsized challenge. At this juncture, the need for a systematic study on the trends, risks, and implications of cybercrimes as well as other emerging trends on social network is becoming ever more pressing.

This publication is highly relevant to not only academic researchers on cybercrime; it also offers visionary assessments for those in arenas potentially affected by technology criminals and for those decision-makers who would like to deepen their understanding of the issues. Its balanced blend of updated perspectives, inclusion of different response tactics used in countries with different cultures, and the contributors' observations on how cyber terrorism, organized cybercrime, and technology-enabled financial crimes will further evolve, provide readers with solid discussions and sharp insights on the issues. Throughout the chapters, there is a discussion on risk profiles in different regions and their varied models of response through prevention and enforcement activities across East and West. Themes that emerged from the conference in 2013, including cybercrime versus national security, cyber-crowdsourcing versus privacy, hacktivism and human rights versus government transparency and openness are well discussed in depth in essays of this book, and they are of the utmost significance in today's evolving cybercrime landscape.

Contemporary risks explored in this volume including threats to national security, spam, cyber terrorism, fraud, and sexually explicit content are challenges common to countries regardless of location. Some call cybercrime a new industry – not entirely an exaggeration, for the revenue criminals are making from it. In 2011, Britain's Office of Cyber Security and Information Assurance reported that the total cost of cybercrime to the British economy was £27 billion (or USD$43.5 billion) per year (Detica and Office of Cyber Security and Information Assurance 2011). As security software company McAfee puts it, cybercrime is a growth industry with high returns and low risks. Their estimation in 2014 reported an approximate annual cost of cybercrimes to the global economy of more than USD$400 billion (McAfee and Center for Strategic and International Studies 2014).

In the past years, one of the trends that captured the attention of law enforcement agencies and researchers alike was the formation of cyber-criminal alliances and networks that switch roles depending on the opportunity of crime. Some are very sophisticated in that they employ a decentralized global C2C (criminal to criminal) model similar to the traditional B2B business model. Centralized points of origin are often absent with complex evasion techniques, increasing the difficulty for digital forensics and law enforcement across borders. Governments, in response, have significantly stepped up protection of critical infrastructures, as some of the alliances have already proven their ability to overcome cyber defense.

Chapter 9 of this book covers another alarming trend, which is the gradual shift of motives for committing cybercrimes. Extremist organizations such as ISIS have displayed meticulousness in the use of various social media, especially Twitter, to spread its message and lower the cost of recruitment. Hacktivism, such as that employed by Anonymous, as explored in Chapter 17, has become a double-edged sword. Often referred to as politically motivated hacking, hacktivism is becoming common around the world, as Chapter 16 lucidly illustrates. These essays enable readers to reflect on the complexity of values and principles in today's digital environment.

Meanwhile, whistleblowers – the insiders disclosing to the public the immoral, improper, or illegal practices in an organization or government – are protected by laws in some countries. However, as the Snowden and WikiLeaks incidents have shown, when the subjects of the whistleblowers are governments themselves, the issues have become much more complicated, and any previously available protection by law to whistleblowers is rendered useless. It is now harder and harder to draw the line between citizens' rights to personal privacy and national security and combating cybercrime.

I believe that laws and regulations must be technology-neutral as much as possible. Criminal justice and forensic responses to cybercrime require heightened technical capabilities in law enforcement operating in an opaque arena, yet expectations from citizens have grown, and they expect a much higher degree of transparency. These emerging conflicting priorities and expectations have certainly risen to the surface of public attention. People do not want to be hacked by criminals, but they certainly do not want to be hacked by their own government either.

Reading this book reconfirms my belief that even in light of huge cybercrime risks, members of the public must remain steadfast in requesting oversight, transparency, and respect for all individuals' privacy. The Internet is just a medium, and we should not shoot the messenger because of the messages carried.

Placed at the intersection of East and West, both geographically and socially, Hong Kong is highly prone to organized, sophisticated, and highly motivated cybercriminal groups pursuing monetary or political goals, while a high level of confidence and security is imperative to support business operations and the city's reputation as an international financial center. This publication delves into practical security issues facing industries and the experience in retail payment and banking in Chapters 11 and 12, and I am sure the analysis of Hong Kong's experience to mitigate cybersecurity risks would be valuable to readers.

This book offers great help in addressing crucial issues of cybersecurity in a timely and structured manner. Is it high time to rethink if we should automatically accept anything in the name of security? What do we do about state-sponsored cybercriminals? These essays shed light on such questions, and I am sure that better understanding of these will allow us to formulate better response strategies and consider necessary legal frameworks to cope with cybersecurity risks and strive to maintain a free and open Internet.

Again, I wish to express my heartfelt gratitude to all the authors and editors in putting together this important book.

Charles Mok
Legislative Councillor (Information Technology)
Hong Kong Special Administrative Region
March 2015

Bibliography

Detica and Office of Cyber Security and Information Assurance (2011) *The Cost of Cyber Crime*, London: Cabinet Office. http://www.cabinetoffice.gov.uk/resource-library/cost-of-cyber-crime (accessed 17 March 2015).

McAfee and Center for Strategic and International Studies (2014) *Net Losses: Estimating the Global Cost of Cybercrime Economic Impact of Cybercrime II*, Washington, DC: Center for Strategic and International Studies.

Acknowledgments

We are grateful to many individuals and organizations for assisting with the preparation of this volume. In particular, we appreciate the financial contributions made by our respective employers who allowed us time in which to edit the chapters and attend to the many tasks that had to be undertaken to finalize the publication. These include the following: the Australian Institute of Criminology, the City University of Hong Kong, and the Asia Pacific Association of Technology and Society (APATAS). We are also grateful to the various organizations that were willing to allow us to republish material from previous works. Finally, we greatly appreciate the professional assistance of the editorial staff of Palgrave Macmillan.

Notes on Contributors

Editors

Ray Chak-Chung Cheung is an assistant professor in the Department of Electronic Engineering at City University of Hong Kong and with Digital Systems Lab. He holds B.Eng. (Hons) and M.Phil. degrees in computer engineering and computer science and engineering from The Chinese University of Hong Kong (CUHK) and a PhD in computing from Imperial College London. He has worked in industry and at Stanford University, UCLA, and Princeton in the United States and the Department of Computer Science and Engineering at CUHK. His current research interests include cryptographic hardware designs and design exploration of system-on-chip (SoC) designs and embedded system designs.

Laurie Yiu-Chung Lau is the chairman of the Asia Pacific Association of Technology and Society (APATAS). He is an entrepreneur and philanthropist in Hong Kong with a PhD from the Faculty of Humanities, Law and Social Sciences at the University of Glamorgan. He has wide-ranging experience in research on policing Internet-related crime, technology, and society. He is also the director of Momentous Asia Travel and Events Co. Ltd. and is actively involved in mentoring students at Wu Yee Sun College and as Honorary Research Fellow of the Pearl Delta Social Research Centre at the Chinese University of Hong Kong.

Russell G. Smith is Principal Criminologist at the Australian Institute of Criminology in Canberra, Australia. He has qualifications in law, psychology, and criminology from the University of Melbourne and a PhD from King's College London. He practiced as a lawyer before becoming a criminology lecturer at the University of Melbourne. He then took up a position at the Australian Institute of Criminology where, in addition to being principal criminologist, he is the head of the Transnational, Organised and Cyber Crime Program. He has published extensively on aspects of cybercrime, fraud control, and professional regulation with over 150 publications. He is the president of the Asia Pacific Association of Technology and Society (APATAS).

Contributors

Mamoun Alazab is Lecturer in Cyber Security in the Department of Policing, Intelligence and Counter Terrorism at Macquarie University in Sydney, Australia. He completed his PhD in IT security from the Federation University of Australia and is a computer security researcher and practitioner with industry, academic, teaching, and research experience in the areas of cybersecurity and cybercrime. He has worked closely with government and industry on many projects, including projects for IBM, the Australian Federal Police (AFP), the Australian Communications and Media Authority, Westpac, and the Attorney General's Department.

Martin Bouchard is an assistant professor in the School of Criminology at Simon Fraser University in British Columbia, Canada. He obtained his PhD from University of Montreal in 2006. His research interests include the organization and dynamics of illegal markets, social capital theory, social networks, methodologies to estimate the size of illegal populations, criminal careers, organized crime, and terrorism. He has edited two books and is the author of numerous articles in peer-reviewed journals, including *Justice Quarterly, Journal of Research in Crime and Delinquency, Journal of Criminal Justice,* and *Journal of Quantitative Criminology.* Dr. Bouchard is the recipient of the 2013 Dean of Graduate Studies Award for Excellence in Graduate Supervision.

Roderic Broadhurst is Professor of Criminology at the Research School of Social Sciences, Australian National University in Canberra. He has published research on cybercrime, recidivism, crime victims, organized crime, homicide, and other violent crime. His latest book, *Violence and the Civilizing Process in Cambodia,* will appear in 2015.

Barry Cartwright is a senior lecturer in the School of Criminology at Simon Fraser University in Canada and a member of the university's International CyberCrime Research Centre. He was previously employed by the solicitor general of Canada, and although not a lawyer, he was for many years a managing partner of a Canadian immigration law firm. He has a BA and MA in sociology and a PhD in criminology. His current areas of specialization include cyber research, cyber bullying, cyber terrorism, and digital copyright law.

Darrell Chan is an assistant researcher at the People's Public Security University of China, where he holds a master's in criminology. He has also been awarded a Certificate of the Legal Profession Qualification. He was a visiting fellow at the Australian National University. He is majoring in criminology with a focus on prevention and countermeasures on cybercrime as well as legislation and international cooperation. He is also doing research on organized crime. Since 2005 he has published articles on cybercrime, victim research, and crime prevention.

Lennon Y. C. Chang is a lecturer in the School of Social Sciences at Monash University, Australia. He was awarded his PhD by the Australian National University in 2010 and then worked as an Assistant Professor of Criminology in the Department of Applied Social Sciences at the City University of Hong Kong from 2011 to 2014. Dr. Chang has conducted research into crime and governance of cyberspace, particularly in the Asia-Pacific region. His book *Cybercrime in the Greater China Region: Regulatory Responses and Crime Prevention* (2012) is about the nature and range of responses to cybercrime between China and Taiwan.

Yachi Chiang is an assistant professor at the National Taipei University of Technology Intellectual Property Graduate Institute in Taipei, Taiwan. She is also the director of the NTUT patent licensing and technology transfer center and the NTUT Representative of Taiwan NCP for EU Horizon 2020 Project. She holds a law degree from National Taiwan University, an LLM in cyber law from Leeds University, and a PhD from Durham University. After working in the Taiwan Legislative Yuan and law firms, in 2012, she launched her academic career as an assistant professor in Shih-Hsin University and then moved to National Taipei University of Technology.

Garth Davies is an associate professor in the School of Criminology at Simon Fraser University in Canada. He has a BA and MA from Simon Fraser University and a PhD from Rutgers University. He conducts research into cyber terrorism, terrorism, communities and crime, housing and crime, policing, criminological theory, statistical analyses, and research methods.

Richard Frank is an assistant professor and associate director of the International Cybercrime Centre at Simon Fraser University in Canada. He completed a PhD in computing science and another PhD in criminology at Simon Fraser University. His main research interest is computational criminology – the application of computer solutions to model or solve crime problems. Cybercrime is another area where his two domains mix. Specifically he's interested in hackers and security issues, such as online terrorism and warfare. Dr. Frank has recent publications at top-level datamining outlets such as in Knowledge Discovery in Databases and security conferences such as Intelligence and Security Informatics.

Peter Grabosky is a professor emeritus at the Australian National University in Canberra and a fellow of the Academy of the Social Sciences in Australia. He has received the Sellin-Glueck Award of the American Society of Criminology for contributions to comparative and international criminology; the Prix Hermann Mannheim from the Centre International de Criminologie Comparée, Université de Montréal for contributions to the development of comparative criminology; and the Gilbert Geis Lifetime Achievement Award from the US National White Collar Crime Center and the White Collar Crime Research Consortium in recognition of outstanding professional contributions in the area of white collar crime.

Leung Ka Ho (Andy) is a research assistant in the Department of Applied Social Science at City University of Hong Kong. He was awarded his BSc in social sciences in the same department, majoring in criminology in 2011. He worked as a Project Officer in Excel Life and Development Services Centre, mainly on youth work training. He worked as a research assistant on a cyber bullying project supervised by Professor Dennis Wong and has been working on projects of cyber vigilantism in Greater China region supervised by Dr. Lennon Chang.

Alice Hutchings is a research associate at the University of Cambridge. She is a criminologist, specializing in cybercrime, and she previously worked at the Australian Institute of Criminology where she researched crime in the cloud, online child sexual offending, communication confidentiality, consumer fraud, identity crime, and computer security risks for small business. Alice's doctoral research examined why computer crime offenders who engage in unauthorized access and fraud begin offending, why they continue, and why they stop. She used a framework of multiple theories of crime, which were integrated to create one theoretical model based on data from current and former offenders.

Lachlan James is a technology entrepreneur with a broad career spanning venture capital, technology commercialization, and IP/IT law. He has qualifications including a master of laws from the University of Cambridge and other qualifications including engineering, law, business, and finance from the University of Melbourne, the Australian National University, and Harvard Business School.

Shu-pui Li is the head of financial and infrastructural development for the Hong Kong Monetary Authority (HKMA). He is responsible for developing and promoting the financial infrastructure of Hong Kong and coordinating with overseas authorities on cross-border financial infrastructure development. Mr. Li has over 20 years' experience in auditing and management consultancy services, technology risk management, banking supervision, and financial infrastructure development. Prior to joining the HKMA, he was the head of Asia Technology Risk Management of Chase Manhattan Bank and the Regional Information Risk Management Officer of JPMorgan Chase in Asia. Mr. Li is a qualified chartered accountant, a qualified chartered information system engineer (CEng), and a certified information systems auditor (CISA). He holds a research master's degree in computer graphics and a bachelors (1st class honors) degree in computing science from the University of Manchester.

Paul Tak Shing Liu is the director of Ritzey Research Institute Limited in Hong Kong. He holds a PhD in electrical engineering from McGill University and degrees in physics and electrical engineering from Lakehead University in Ontario. He is a member of the Institute of Electrical and Electronic Engineers, a fellow of the Hong Kong Institution of Engineers, and a member of the Hong Kong Computer Society. He was manager of Chong Hing Bank Limited for 15 years and was a director of Chong Hing Information Technology Limited and Net Alliance (HK) Limited. He is school manager of Liu Po Shan Memorial College. He specializes in information security and banking and finance applications.

Alana Maurushat is a senior lecturer and the academic director of the Cyberspace Law and Policy Community at the University of New South Wales, where she received her PhD. She is key researcher on the Critical Research Centre "Data to Decision" and a Senior Research Fellow with the Australian Cyber Security Centre. Her most recent book examines legal issues in the disclosure of security vulnerabilities and exploit markets, and forthcoming books examine ethical hacking and botnets. She is on the board of directors of Internet Fraud Watchdog and has presented at many conferences around the world and in the media.

Fernando Miró is Professor of Criminal Law and Criminology at the Miguel Hernandez University of Elche, Spain, and also the director of the Crimina Center dedicated to the study and prevention of crime. He works in different aspects of modern criminal law, the emerging crimes that relate to economics and business, and also law of the new technologies, and is one of the most highly respected and referenced specialists in cybercrime in Spain. His research interests are wide ranging but have mostly focused on issues in victimology, including theories of victimization, cybercrime victimization, and stalking victimization. Professor Miró earned his PhD in criminal justice from the University of Alicante, and he has written 40 books and book chapters and over 33 journal articles.

The Hon. Charles Mok is a member of the Legislative Council of Hong Kong representing the Information and Technology Functional Constituency. He has been serving the information technology and telecommunication industry for more than 20 years. He is the chairman of Professional Commons and co-founder of Internet Society Hong Kong. He is active in public services with a view to upholding the core values of Hong Kong, including democracy, liberty, human rights, rule of law, and integrity. He holds bachelor's and master's of science degrees in computer and electrical engineering from Purdue University in the United States.

Gregor Urbas is Associate Professor of Law at the University of Canberra, Australia, where he teaches cybercrime, criminal law and procedure, and evidence. He is a co-author of *Cyber Criminals on Trial* (2004) and numerous articles and reports on cybercrime, intellectual property enforcement, and criminal justice issues. He previously held positions at the Australian National University, the Australian Institute of Criminology, and the Law Council of Australia, and he is an adjunct associate professor at the ANU College of Law and at Simon Fraser University in Vancouver, Canada.

Dawei Wang is Professor of Criminology and Director of the Police Office Education and Training Research Center at the People's Public Security University of China. He holds a PhD in sociology from Tsinghua University. He is a lifetime member of the World Society of Victimology, a standing member of the Council of Chinese Society of Criminology

(CSC), and the deputy president of CSC. He has published six books, written and translated 47 papers which focused on issues of organized crime and victimology, as well as being the author of 16 jointly written or edited books.

George Weir is Lecturer in Computer and Information Sciences at the University of Strathclyde in Glasgow, Scotland. He holds academic qualifications from three Scottish universities, including a PhD from the University of Edinburgh. In addition to teaching and research supervision, he has published extensively in the areas of cybercrime, security, readability, and corpus linguistics. He is a member of the Forensics and Cybercrime Specialist Group, the Association of Computing Machinery, the British Computer Society, IEEE, and the Higher Education Academy. He is also a fellow of the Winston Churchill Memorial Trust and a fellow of the Royal Society of Arts.

1

Introduction: Cybercrime Risks and Responses – Eastern and Western Perspectives

Russell G. Smith, Ray Chak-Chung Cheung, and Laurie Yiu-Chung Lau

Crime has traditionally been a domestic concern for nation-states, with individual jurisdictions having their own responsibility for policing, prosecution, criminal trial, and the imposition of judicial punishments. Although some conventional crime types, such as maritime piracy, smuggling, and organized crime have always crossed jurisdictional borders and required transnational cooperation for investigation and judicial responses, it was not until the mid-twentieth century that crime became a more globalized phenomenon. As Findlay (1999, p. 2) observed:

> The globalisation of capital from money to the electronic transfer of credit, of transactions of wealth from the exchange of property to info-technology, and the seemingly limitless expanse of immediate and instantaneous global markets, have enabled the transformation of crime beyond people, places and even identifiable crimes.

Findlay goes on to explain that "having said this, our world is still far from the universalised culture in all quarters" and that "globalisation is a transitional state" (1999, p. 2).

Since 1999, however, developments in information and communications technologies (ICT) have, arguably, facilitated the process of globalization markedly and created a world that, in many respects, crosses jurisdictional boundaries freely and without difficulty. Criminals have benefited greatly from developments in ICT in terms of facilitating communication, identifying potential victims of economic crimes, and obtaining information that can be used to facilitate crime. Criminals

1

have also used ICT to counter the efforts of law enforcement to detect and respond to crime, such as through hacking into secure policing networks, encrypting information against law enforcement inspection, and laundering the proceeds of crime electronically.

One important question that organizations such as the United National Office on Drugs and Crime (UNODC) has examined is how different regions of the world have experienced technology-enabled crime and responded to it through prevention and enforcement activities. The use of ICT by transnational and organized crime groups has created considerable concern internationally (Choo and Smith 2008), with the UNODC beginning to develop normative approaches to deal with the problem (UNODC 2013). In the Asia-Pacific region, attention has also been focused on these concerns, with a number of international conferences and events held to share information on how best to respond to the challenges that new technologies have created. Of particular relevance are the Asia Cyber Crime Summits held in Hong Kong in April 2001 and November 2003, whose proceedings were published by Broadhurst (2001) and Broadhurst and Grabosky (2005). More recently, Chang (2013) has examined cybercrime in the greater China region and reviewed various regulatory responses and crime-prevention initiatives in Mainland China, Hong Kong, and Taiwan.

In August 2013, an international conference on cybercrime and computer forensics organized by the City University of Hong Kong and the Chinese University of Hong Kong was held in Hong Kong. Thirty speakers from ten countries across the region, as well as from North America and Europe, examined the most critical issues facing governments, business organizations, and consumers in relation to computer-enabled crime. From this group of policy and business experts, perspectives on cybercrime were obtained that could be said to reflect both Eastern and Western perspectives. A selection of papers presented at this conference form the chapters published in the current volume.

Clearly, many of the issues concerning computer crime victimization and the use of technology to respond to cybercrime are common across nations, be they in the East or in the West. Victimization experiences are, to a large extent, highly similar, whether they concern malware infections, consumer fraud, dissemination of illegal content, or facilitation of terrestrial crimes such as copyright infringement, theft, or stalking. An indication of some areas of difference in cybercrime victimization is apparent from the results of a survey undertaken by the UNODC (2013) in which the most common types of cybercrimes encountered by national police in four principal global regions were assessed. It was found that crimes involving child pornography were more prevalent in police investigations in Western than in Eastern regions, while unauthorized access to networks was more prevalent in Eastern than in Western regions. Computer-related acts involving racism and xenophobia were only considered to be prevalent in police investigations in Asia and Oceania (UNODC 2013, p. 26). Such differences, however, reflect the priorities of regional law enforcement agencies rather than the underlying incidence of different cybercrime types.

The present volume's Eastern and Western perspectives derive more from the location of the authors of individual chapters than the content of any continental differences. Seven come from Greater China, five from Australia, three from Europe, and two from North America. It is apparent that most of the issues being discussed among cybercrime researchers today are common across national borders, with differences only being apparent in the incidence of some forms of cybercrime and the rate of implementation of specific response strategies. For example, the incidence of computer access-related crime appears to be more prevalent in China than in Western nations, while content offending, particularly involving child-exploitation material, is prosecuted more often in the West than in the East (Smith, Grabosky and Urbas 2004).

In relation to responses to cybercrime, there have been marked differences in the roll-out of technologies used to minimize risks. In the case of payment card fraud, Roland Moreno introduced smart cards in France in 1974, while the United States is just beginning to roll out chip-enabled credit cards in 2015, albeit without the use of PINs. Similarly, biometric user authentication systems have been implemented in some countries at a much earlier time than in other countries (see Chapter 2). Arguably, cloud computing will be taken on in all developed nations at much the same time – with all the benefits and risks that that might entail (see Chapter 10 by Alice Hutchings et al. below). These and other developments are examined in the following chapters by academic scholars, business professionals, and government policy-makers from both Eastern and Western countries. Their solutions to the problem, it will be seen, have much in common.

The present volume is organized in four parts. The first begins with an historical review of cybercrime globally and then examines two areas of cybercrime research that raise both methodological and ethical questions as well as substantive issues of access to, and analysis of, information posted online (Chapters 3 and 4). Part II examines six contemporary cybercrime risk topics affecting individuals, business, and government with contributors from both Eastern and Western countries (Chapters 5 to 10). Part III considers a range of industry, criminal justice policy, and forensic responses to cybercrime that have relevance to business and governments globally (Chapters 11 to 14). The final part presents three research-based studies that concern privacy and freedom in cyberspace, particularly in the realm of social media (Chapters 15 to 17). The chapters include current theory and information on cybercrime risks for the future and a wide range of policy and industry responses available to counter the growing threat to information and communications technology infrastructure. Unlike some other publications that have addressed the problem of cybercrime from a purely academic perspective, the present collection includes not only research from government policy analysts but also university-based researchers and theorists and representatives from industry who have agreed to share their perspectives on how some sectors, such as banking and financial services, have sought to address cybercrime risks in recent years.

Some of the principal issues and themes explored in the following chapters are as follows. At the outset, it is important to understand the rapid growth in ICT usage in the region. In their discussion in Chapter 16 of cyber-crowdsourcing, or what is known in China as "human flesh searching," Lennon Chang and Andy Leung cite Internet World Statistics (2015) that there are more than 650 million Internet users in the Greater China Region, most of whom have a presence in online communities such as Golden (HK), HK Discuss (HK), U-wants (HK), PPT (Taiwan), Weibo (China), and Facebook (HK and Taiwan). China Internet Watch (2014) also reports that there were more than 167 million monthly active users on Weibo in 2013, while in Taiwan and Hong Kong, 70 percent of Internet users are on Facebook. Chang and Leung observe in Chapter 16 that these substantial numbers of "netizens" and platforms provide an excellent base for cyber-crowdsourcing activities in the Greater China region.

Of course, similarly high proportions of citizens in Western nations are also online. Internet penetration statistics that show the percentage of the population in various regions who are users of the Internet vary considerably, as shown in Figure 1.1. North America has the highest penetration rate of 87.7 percent of the population, while Asia has the second lowest rate of 34.7 percent. However, in terms of numbers of users, Asia has by far the largest number of Internet users of any region in the world, with 1,386 million users at 30 June 2014, or 45.7 percent of all the world's Internet users (Miniwatts Marketing Group 2015, n.p.).

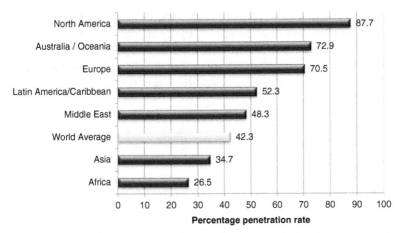

Figure 1.1 World Internet penetration rates by geographical region – Quarter 2, 2014

Miniwatts Marketing Group (2015, n.p), used with permission.

The relevance of these statistics lies in crime opportunity theory, explained by Russell Smith in Chapter 2. Opportunity theory seeks to explain trends in crime rates in terms of changes in the routine activities of everyday life. It is argued that changes in routine activity patterns influence crime rates by affecting the convergence in space and time of three minimal elements of direct-contact predatory violations: the presence of motivated offenders, the availability of suitable targets, and the absence of capable guardians against a violation. Having such a large proportion of the population using the Internet and other electronic devices creates a high-risk environment for crime, particularly in the presence of high financially driven motivations for offending and the difficulties that the Internet presents in terms of effective policing or guardianship. Although victims of cybercrime are located in both Eastern and Western countries, it can be expected that the future will see the highest incidence of victimization, in terms of numbers of victims, in Asia. In addition, such a high rate of ICT usage will inevitably mean that high numbers of individuals based in the Asian region may choose to engage in cybercrime perpetrated against not only their own citizens but individuals from all nations. Motivations for offending are likely to be great in countries with a low GDP but ready access to inexpensive technologies. As a result, the pressure on government and business to implement effective preventive strategies remains paramount and urgent.

The ever-expanding use of ICT in both the East and the West has created opportunities for criminality that were considered to be purely hypothetical over a decade ago. In Grabosky and Smith's (1998) early review, *Crime in the Digital Age*, questions of cyberterrorism, organized cybercrime, misuse of social media, wireless and mobile technological risks, and cloud computing were not on the horizon, but they now occupy a prominent position in policy and industry discussion. Peter Grabosky, in Chapter 5, explores the question of whether organized cybercrime constitutes a national security threat, arguing that the answer depends on one's definitions of "national security" and "organized cybercrime." A response to this comes from Gregor Urbas, in Chapter 13, who presents a clear exposition of the highly technical laws that govern accessorial liability when applied to online organized crime. It is suggested that a more explicit recognition of "virtual presence" is required as the key legal element that could also assist in resolving persistent questions about identifying where a crime was committed.

One of the commonalities of cybercrime across nations is the financial motivations that underlie the vast majority of acts of illegality perpetrated online. Laurie Lau, in Chapter 6, provides a review of a selection of court cases involving Internet banking fraud in Hong Kong and demonstrates how the rule of law has helped Hong Kong achieve its position as an economic hub in Asia despite these online risks. Data from the Australian Cybercrime Pilot Observatory are presented in Chapter 7 to illustrate the important role that spam plays in enabling financial crime online. Alazab and Broadhurst explain some of the new methods of locating new victims of financial crime, such as right-to-left override, spear phishing, and URL shortening, and consider how different forms of social engineering relate to different types of malware and crime. They argue that because the focus of ICT security on perimeter protection is becoming increasingly ineffective, there is a need to refocus crime prevention activities on the *modus operandi* of offenders. Turning to Asia, Darrell Chan and Dawei Wang present the results of their study of cybercrime perpetrators in China in Chapter 14, examining their typical characteristics and considering the extent to which organized cybercrime groups are involved in cybercrime in China.

The problem of financially motivated cybercrime is present not only in Western nations but also in China. Various attempts have been made to estimate the costs of cybercrime, and it is generally agreed that costs to government, business, and individuals total many billions of dollars annually. In 2014, research by the Center for Strategic and International

Studies for McAfee estimated that cybercrime cost 0.64 percent of GDP in the United States, 0.63 percent of GDP in China, 0.41 percent of GDP in the European Union, 0.16 percent of GDP in the UK, and 0.08 percent of GDP in Australia (McAfee and Center for Strategic and International Studies 2014, p. 21). The total annual cost to the global economy from cybercrime was estimated to be more than US$400 billion. A conservative estimate was US$375 billion in losses, while the maximum could be as much as $575 billion (McAfee and Center for Strategic and International Studies 2014, p. 2).

Some regions experience higher rates of cyber-victimization than others, with the UNODC (2013, p. 30) explaining this as follows:

> Differences in average reported loss across countries are likely due to a number of factors, including the type of cybercrime victimization, the effectiveness of cybersecurity measures, and the extent of victim use of the internet for online banking or payments.

In order to counter the efforts of cybercriminals, banks, in particular, undertake extensive prevention measures, costing substantial amounts each year. Chapters 11 and 12 provide an indication of how the banking industry is responding to new and emerging risks of cybercrime in Hong Kong. Paul Liu provides an historical overview on Hong Kong's banking information technology security practices and discusses how banks in Hong Kong have developed their information technology security in line with advances in modern technology in the Hong Kong banking market. In Chapter 11, Shu-Pui Li, Head of Financial Infrastructure Development with the Hong Kong Monetary Authority, considers how, through various initiatives, the Hong Kong Monetary Authority (HKMA) has enhanced Hong Kong's retail payment infrastructure, including the development of e-bills and e-cheques and the introduction of a new regulatory framework for stored-value facilities. Although these security measures are laudable, financially motivated cybercrime continues apace, leading researchers at the University of Cambridge Computer Laboratory to observe that:

> The straightforward conclusion to draw on the basis of the comparative figures collected in this study is that we should perhaps spend less in anticipation of computer crime (on antivirus, firewalls etc.) but we should certainly spend an awful lot more on catching and punishing the perpetrators (Anderson et al. 2012, p. 26).

In Chapter 14, Darrell Chan and DaWei Wang use the principles of traditional crime prevention to develop a comprehensive framework of countermeasures to prevent and control cybercrime in China, particularly focusing on situational crime prevention approaches with an emphasis on victim-oriented prevention. Arguably, this, in conjunction with criminal justice approaches, provides a more effective response than simply target hardening, which has proved to be elusive in a global environment in which technology is effective in placing barriers in the path of conventional law enforcement interventions.

Part IV of the book explores a number of questions relating to privacy and freedom in the online world. Yachi Chiang, in Chapter 13, uses Facebook to illustrate the risks that face users of social networking sites relating to infringement of their privacy. Since commencing business in 2004, Facebook has adopted a variety of privacy setting models concerning protection of personal information. By comparing relevant regulations in the United States, the European Union, and Taiwan, the appropriateness and feasibility of protecting users' privacy is explored. Lennon Chang and Andy Leung then examine the phenomenon of cyber crowdsourcing, known as "human flesh searching" in China, and how this can be used to assist in policing cyberspace. The authors also explore the counterproductive consequences of vigilantism in the online world and the potential and actual harms that have been caused in the Greater China Region.

The collection ends with an imaginative and spirited exploration of civil rights activism in the digital age. Recalling the experiences of Leni Reifenstahl, the artist who became embroiled in controversy when she produced propaganda for the Nazis in the 1940s, Alana Maurushat explores the world of hacktivism and whistleblowing and its impact on information transparency governance. She poses the question of whether hacktivists should be reviled or revered in the digital age.

Throughout the volume we see commonalities and divergences between policy and legislative and industry responses to cybercrime in a selection of Eastern and Western nations. Questions of freedom of expression, personal liberty, and unfettered access to content are pitted against the needs of business and government to maintain control of online behavior and to permit lawful access to the Internet for policing and national security purposes. Each country maintains its own balance in these competing interests, but the globalization of the online world has meant that increasingly, international normative approaches have become necessary. Arguably, the time has now arrived for the creation of a global Cybercrime Convention that will cater for both Eastern and

Western interests in cyberspace alike. Unfortunately, the diversity of societies in these two regions in terms of politics, economy, technology, religion, culture, and affluence will make the task for the United Nations in achieving consensus as to the content of a Cybercrime Convention challenging. It is, however, a challenge that is well worth pursuing.

Bibliography

Anderson, R. et al. (2012) "Measuring the Cost of Cybercrime," *11th Annual Workshop on the Economics of Information Security*, Berlin: WEIS 2012, 25–26 June.

Broadhurst, R. (2001) *Proceedings of the Asia Cybercrime Summit*, Hong Kong: The University of Hong Kong.

Broadhurst, R. and Grabosky, P. (2005) *Cybercrime: The Challenge in Asia*, Hong Kong: Hong Kong University Press.

Chang, L. Y. C. (2013) *Cybercrime in the Greater China Region: Regulatory Responses and Crime Prevention Across the Taiwan Strait*, Cheltenham, UK: Edward Elgar.

China Internet Watch (2014) "Weibo had 167M Monthly Active Users in Q3 2014," http://www.chinainternetwatch.com/10735/weibo-q3-2014 (accessed 28 February 2015).

Choo, K-K. R. and Smith, R. G. (2008) "Criminal Exploitation of Online Systems by Organised Crime Groups," *Asian Criminology* 3, 37–59.

Findlay, M. (1999) *The Globalisation of Crime: Understanding Transitional Relationships in Context*, Cambridge, UK: Cambridge University Press.

Grabosky, P. N. and Smith, R. G. (1998) *Crime in the Digital Age: Controlling Telecommunications and Cyberspace Illegalities*, Sydney: Federation Press/New Brunswick: Transaction Publishers.

McAfee and Center for Strategic and International Studies (2014). *Net Losses: Estimating the Global Cost of Cybercrime – Economic Impact of Cybercrime II*, Santa Clara, CA: McAfee and Center for Strategic and International Studies.

Miniwatts Marketing Group (2015) "Internet World Stats: Usage and Population Statistics," http://www.internetworldstats.com/stats.htm (accessed 28 February 2015).

Smith, R. G., Grabosky, P. N. and Urbas, G. F. (2004) *Cyber Criminals on Trial*, Cambridge, UK: Cambridge University Press.

United Nations Office on Drugs and Crime (UNODC) (2013) *Comprehensive Study on Cybercrime*, New York: UNODC.

Part I

Understanding Cybercrime Through Research

2

Trajectories of Cybercrime

Russell G. Smith

Acknowledgment

This chapter is a condensed and revised version of the author's earlier work (Smith 2010b), "The Development of Cybercrime: An Opportunity Theory Approach," originally published in *Crime Over Time*, edited by Robyn Lincoln and Shirleene Robinson (Cambridge Scholars Publishing, Newcastle upon Tyne). It is included in the current collection with permission.

Introduction

In 1979, Cohen and Felson (1979, p. 589) raised for consideration a theory to explain trends in crime rates in terms of changes in the routine activities of everyday life. They argued that the structure of ordinary daily activities influences criminal opportunities, especially for so-called "direct-contact predatory violations." Structural changes in routine activity patterns were said to influence crime rates by affecting the convergence in space and time of the three minimal elements of direct-contact predatory violations: the presence of motivated offenders, the availability of suitable targets, and the absence of capable guardians against a violation.

Some 20 years later, Felson and Clarke (1998, p. 22) outlined ten principles of opportunity theory, one of which argued that social and technological changes produce new crime opportunities. Although not all technological changes result in new opportunities for crime, those that correspond with people's motivations for illegality and are not subject to capable guardianship are likely to lead to new or adapted forms of crime. Felson and Clarke (1998) described the life cycle of crime opportunities in connection with mass-produced consumer goods as involving four stages: innovation, growth, mass market, and saturation. Adapting

13

their approach to the world of computers, a crime opportunity structure would operate as follows:

> In the *innovation* stage, the product is sold to a special group of consumers. It may be expensive, difficult to use, relatively heavy and awkward. That explains why early computers and more recently older-style mobile phones were not likely to be stolen. In the *growth* stage, products become easier to use, cheaper to buy, lighter and less awkward to carry. More people discovered how to use them and wanted one, and thefts therefore accelerated. In the realm of computing, an example is the development of laptop computers which were small, valuable and desirable. In the *mass market* stage, the product gains further in appeal. More units are sold and theft becomes endemic, such as we have seen with mobile phones. By the *saturation* stage, most people who really want the product have it, and thefts decline. For example, hand calculators sell for a few dollars and are mostly safe on your desk with the door open. Many products that once fed the crime wave are now in the saturation stage, offering little incentive to theft. As innovations occur, new products enter the same cycle. In the world of computing the latest innovations such as iPods and Tablets remain expensive, desirable and are likely to be compromised or stolen (adapted from Felson and Clarke 1998, p. 22).

Approaches to crime prevention have targeted each of the essential components of opportunity theory: reducing the motivations for individuals to commit crime; reducing available targets for the commission of crime; and enhancing the role of police, regulators, and crime prevention agencies in ensuring that crime is prevented or detected. These, of course, focus on the periods prior to saturation when crime is endemic. Thereafter, the market should regulate itself, and (we hope) crime will reduce as products lose their attraction and value.

In cyberspace, we have an ideal criminogenic environment as there are abundant targets and opportunities, highly motivated offenders, and, until very recently, not a great deal of regulation and enforcement. Although opportunity theory was originally devised to understand predatory property crime, it is possible to apply it, with some adaptations, to the development of crime risks in cyberspace.

The present chapter seeks to explore the goodness-of-fit of this criminological approach to understanding the full range of crimes affecting or making use of information and communication technologies (ICTs) that have developed since their inception. In terms of opportunities for crime, we can recall the comment of Willie Sutton, the notorious American bank robber of the 1950s, who was once asked why he persisted in robbing banks. "Because that's where the money is," he is said to have replied (cited by Grabosky and Smith 1998, p. 1). In cyberspace, opportunities abound and these have largely followed the development and introduction of new digital technologies. In the words of Grabosky and Smith (1998, p. 1):

> since Wheatstone and Cooke first patented their system of communication by the means of electromagnetic impulses carried over wires in 1837, crimes have been committed either through the misuse of telecommunications equipment, or against telecommunications equipment. Every technological development has provided a new opportunity for criminality which has often been exploited.

In terms of the motivations behind the commission of technology-enabled crime, there have also been notable changes over the last century. The earliest computer criminals, in the 1970s, were often motivated by curiosity and the challenge of defeating security systems. For example, the famous "phreaker" John Draper, "Cap'n Crunch," discovered be could disable the billing systems on telephone systems by blowing into the phone receiver with a toy whistle. In the 1980s, offenders were attracted by the fame and notoriety they could achieve when they defeated the security measures used by government agencies or banks. Kevin Mitnick, an infamous hacker in the 1980s, committed a series of offenses in an attempt to defraud large corporations until he was arrested in 1995. He was motivated principally through a desire to enhance his reputation among the online community (details of these and other cases referred to below are contained in Smith, Grabosky and Urbas 2004, Appendix A).

Personal motivations have always been evident in some types of cybercrime. For example, sexually motivated stalking, online bullying, or sharing of obscene materials are all founded in personal and individual life circumstances. Revenge has also motivated attacks by disgruntled employees seeking to deface websites or to steal or contaminate business data. In 1994, for example, a man in Sydney, Australia, who had been refused a job with an Internet service provider, defaced its website and disclosed client account information to journalists as an act of revenge ([1999] NSWCCA 69). Similarly, in 2000, a man who had lost his job with a local council in Brisbane, Queensland, caused raw sewage to overflow on the Sunshine Coast by using radio transmissions to alter the operation of council sewage pump stations ([2002] QCA 164). Other individual motivations include those who seek to change government policies for religious or other reasons who may become involved in acts of cyber terrorism (Denning 2010).

Perhaps the most important change has related to the motivations underlying cybercrime. There has been a transition from curiosity and status-driven activities to those that are now predominantly financially motivated and occur in a much more organized and systematic manner. Organized crime groups in the twenty-first century have found that cyberspace provides a rich and lucrative source of income with far fewer risks of violence, prosecution, or punishment. The ongoing risk areas include phishing attacks, identity crimes, payment system fraud, corporate extortion, and money laundering.

Finally, in terms of regulation, there is a wide diversity of stakeholders with an interest in controlling cyberspace, although their efforts continue to be dislocated and uncoordinated, making success rare and difficult to achieve. The capable guardians of cyberspace include computer users; local and international law enforcement; financial and business regulators; government policy and legislative agencies at local, national, and international levels; information technology industry organizations including telecommunications carriers, Internet service providers, and hardware and software manufacturers; small and large business associations and their members; and a range of public interest groups including the media, computer user groups, and consumer protection agencies.

The present chapter explores the criminal opportunities created by new technologies and how they could have been avoided had the lessons of the past been applied to preventing new adaptations of the crimes of the past. In many respects the motivations underlying the most recent manifestations of cybercrime are the same as their original precursors. In addition, preventive strategies have often been identified but not fully implemented or simply ignored, thus allowing similar crime typologies to be repeated unchecked. We would be well advised to recall the observation of George Santayana, in *The Life of Reason* (1905, p. 284), that "those who cannot remember the past are condemned to repeat it." At the outset, however, we need to explore the technological changes that have created opportunities for cybercrime.

Technological developments

Digital technologies lie at the heart of cybercrime and these include computers, communication technologies, and networked services. The term "cyberspace" was first coined by William Gibson in his novel *Neuromancer* (1984) to describe a high-tech society in which people inhabit a virtual world divorced from terrestrial life. It has been used since then in a wide range of contexts to describe almost anything to do with computers, communications systems, the Internet, or, indeed, life in the twenty-first century. Although the present chapter employs the concept of cybercrime, its scope extends to all the crime risks that have arisen from technological changes in information and communications technologies, extending from telephony, through word processing, to the most recent forms of networked wireless and cloud computing.

Figure 2.1 provides a chronological representation of developments in ICT over the preceding 177 years. Although the indicated year of onset may be open to some debate, and the timescale provided is not linear (in order to accommodate the information simply), Figure 2.1 provides an indication of the key innovations that have spanned the diverse fields of communications and computing. Of course, not every innovation in ICT has been represented – just the principal developments that have relevance to our discussion of criminal misuse.

This is not the place to provide a complete history of telecommunications and computing as others have done so fully and clearly already

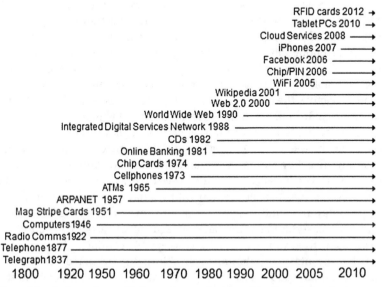

Figure 2.1 A chronology of information and communications technologies

(see Grabosky and Smith 1998 for a review of how communications technologies developed, and Curran 2010 for a recent review of how the Internet evolved). Some of the more important telecommunications milestones include the development of electromagnetic communications by Wheatstone and Cooke in 1837 and the introduction of telephony in 1877. Wireless radio communications were introduced in the 1920s, and telephony services improved as cable and later satellite networks extended their reach. Telex systems and the introduction of subscriber trunk dialling (STD) in the 1950s further improved services, and the 1980s saw international direct dialling (IDD) and Integrated Digital Services Network (IDSN) commence operation, which enabled the transmission of combinations of voice, text, image, and data. Telephone exchanges and complex switching systems enabled calls to be connected electronically using line concentration to maximize outputs. The 1980s saw the introduction of cellular mobile telephone services, initially using an analogue network, which was replaced by digital networks in the 1990s.

Parallel with developments in communications technologies came the introduction of computing in the 1940s and the creation of networked computing following the invention of the ARPANET by the Advanced Research Projects Agency in the United States in the 1950s. Developments in computing and microprocessors also enabled electronic payment systems to be created, including magnetic stripe bank cards in the 1950s and smart cards developed by Roland Moreno in France in 1974. Microprocessing also led to smaller computing machines being created, including cell phones in the 1980s and personal digital assistants (PDAs) and smartphones after 2000. Allied with these hardware developments were improvements and diversification in software that made computing accessible and easy to use. As prices decreased and devices became more attractive, the profile of users changed from those in academic and scientific environments to the general public as a whole.

Throughout the history of ICT, governments have sought to achieve a balance between encouraging the development of technologies and ensuring that they are not being used for illegal or improper purposes. Often, however, instances of misuse have occurred soon after new technologies have been created, when controls have not been put in place. As Grabosky and Smith (1998, p. 14) noted over a decade ago, "the task for policy makers is to ensure that new regulatory strategies operate effectively so as to govern but not stifle acceptable conduct. The challenge which confronts these policy makers resides in the fact that developments in telecommunications technology occur at a pace which is beyond the capacity of policy processes to keep abreast of them."

Developments in cybercrime

Cybercrime also has a lengthy history – that extends back to a time before computers were created. The link lies in the ways in which communications technologies have been compromised for personal and financial gain. Wall (2010, pp. 94–8) has identified three "generations" of cybercrime. The first generation of low-end cybercrime involved the use of mainframe computers and their operating systems to assist traditional forms of offending such as theft of funds or information. The second generation involved offending across computer networks, such as hacking and cracking activities. The third generation of crimes, Wall argues, were wholly mediated by technology and were truly distributed and automated, such as occurs in the dissemination of spam using botnets.

In 1998, Grabosky and Smith (1998) provided a diagrammatic representation of when the various forms of cybercrime that they considered first achieved significance. They argued that "localization of the introduction of each form of offending in history is not able to be isolated with great precision but an indication can be given of which forms of illegality achieved significance earlier rather than later this century and which are of most recent origin" (1998, p. 213). Of course, developments in cybercrime have been extensive since 1998, but it is possible to extend Grabosky and Smith's original chart to present the chronology of the onset of the main types of cybercrime from their inception to the present time. Figure 2.2 provides an updated and extended chronological representation of developments in cybercrime. Once again, the indicated year of onset may be open to some debate, and the time line provided is also not to scale. Finally, not every subcategory of cybercrime is represented – just the principal types that have had the most impact on society.

In reviewing the history of cybercrime, the following discussion will assess how the principles of opportunity theory might contribute to explaining its development, adaptation, and continuation.

Telephony risks

Arguably, the foundations of communications-related crime lie in Oliver Cromwell's era in the seventeenth century when a state monopoly was established over the delivery of letters, expressly so that they could be intercepted to protect the interests of the state (Fitzgerald and Leopold 1987, p. 112). The earliest example of cybercrime, in the electronic sense, involved the illegal interception of telegraph communications that occurred in the mid-nineteenth century.

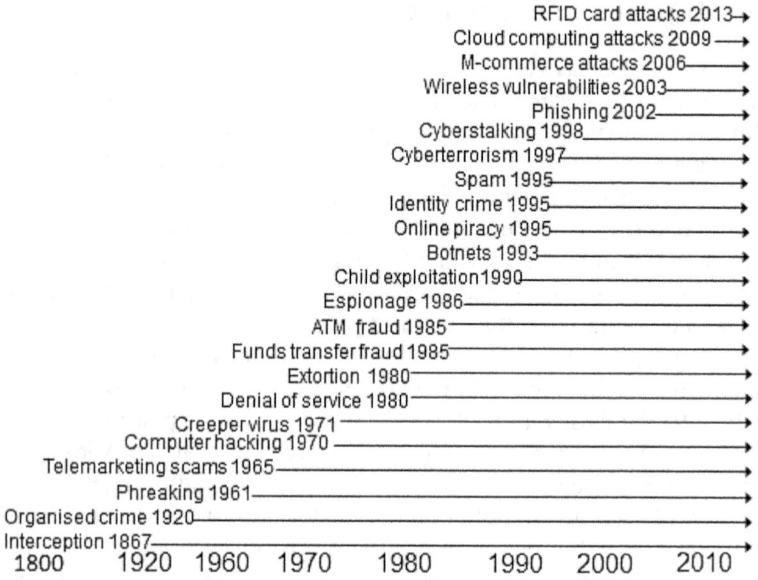

RFID card attacks 2013→
Cloud computing attacks 2009 —→
M-commerce attacks 2006———→
Wireless vulnerabilities 2003————→
Phishing 2002————→
Cyberstalking 1998_____→
Cyberterrorism 1997————————→
Spam 1995————————→
Identity crime 1995————————→
Online piracy 1995————————→
Botnets 1993————————→
Child exploitation 1990————————→
Espionage 1986————————→
ATM fraud 1985————————→
Funds transfer fraud 1985————————→
Extortion 1980————————→
Denial of service 1980————————→
Creeper virus 1971————————→
Computer hacking 1970 ————————→
Telemarketing scams 1965————————→
Phreaking 1961————————→
Organised crime 1920————————→
Interception 1867————————→

| 1800 | 1920 | 1960 | 1970 | 1980 | 1990 | 2000 | 2010 |

Figure 2.2   A chronology of cybercrimes

In 1867, a Wall Street stock broker collaborated with Western Union telegraph operators to intercept telegraph dispatches sent to Eastern newspapers by their correspondents in the West. The intercepted messages were then replaced by counterfeit ones which reported bankruptcies and other financial disasters supposedly befalling companies whose stock was traded on the New York Stock Exchange. When the share prices were driven down, the wiretappers then purchased their victims' stock (O'Toole 1978, p. 97).

The nineteenth century also saw the onset of the first case of electronic piracy. Charles Dickens sued for copyright infringement when his serialized best-sellers were transmitted on a daily basis to the United States, using the trans-Atlantic telegraph. They were then reprinted for American readers without seeking his permission or paying him any royalties (Vaidhynathan 2003). With the development of modern computing, electronic products, such as songs and movies, can all be copied and reverse engineered more quickly and easily than in the past. Information piracy now covers the full range of cybercrime types as computer hardware and software can be copied without permission, and computers can be used to facilitate acts of copying themselves.

In the early years of the twentieth century, offenders found that they could use telephones to organize their criminal activities. One example concerned Al Capone, who used early telephones in the 1920s and 1930s to organize his rackets involving illegal gambling, bootlegging, and prostitution. More recently, a range of organized crime groups continue to make use of ICT. Traditional organized criminal groups, such as the Japanese Yakuza, Asian triads, and Eastern European gangs, have been using ICT to generate funds in connection with software piracy, plastic card fraud, and malware attacks, while other organized cybercriminal groups such as Shadowcrew, DrinkOrDie, Rock-Phish, BotMaster, and Mpack have met in cyberspace with common objectives in connection with activities such as child exploitation, underground malware markets, identity crime, and denial of service attacks. Finally, ideologically or politically motivated groups such as some terrorist organizations have used ICT to raise funds for religious or political change and have been involved in online fraud and money laundering (Choo and Smith 2008).

The 1960s saw a rapid expansion in the use of telephony services that corresponded with attempts at obtaining services for free through "phreaking." The Bell Telephone Company reported the first case of theft of a long-distance telephone service in 1961 when a local office manager discovered some inordinately lengthy calls to an out-of-area directory information number. Since then, criminals have been adept in compromising telecommunications billing systems for gain. Smith (1996) identified over 40 different ways in which services have been stolen since the 1960s when the original "phreakers" started carrying out their activities. Some of the problems related to cloning mobile phones, misusing premium call services, stealing prepaid services, PABX fraud, and diverting calls to international numbers. Internet services and cable TV services can also be stolen.

Theft of telecommunications services provides an excellent example of crime displacement. As each new way of obtaining services for free was closed off from within the industry, new techniques were devised to achieve the same result only through the use of different technological means. In the jargon of the time, the "Blue Box" (switching system bypass) gave way to the "Black Box" (meter inhibition), which gave way to the "Red Box" (coin drop simulation), all within the space of ten years or so (see Clough and Mungo 1992).

Similar displacement effects continue today, particularly concerning malware attacks in which offenders closely monitor the pursuits of anti-malware providers and develop new forms of malware that cannot be detected. Rogue anti-malware programs have also increased considerably in recent years, with PandaLabs in the United States reporting a substantial increase in the creation of rogueware, with 122,335 variations being recorded in the month of December 2009 alone (Anti-Phishing Working Group 2010). "During the first quarter of 2014, PandaLabs gathered 4.8 million new malware samples, with its library reaching 15 million new malware samples in total" (Anti-Phishing Working Group 2014, p. 8).

Computing risks

The 1970s saw the development of a range of activities in which computer networks were compromised, access was obtained illegally, and data were damaged maliciously. Colloquial terms developed including "hacking" (using ingenuity to refine existing computer programs), "cracking" (intruding on computer systems), "vandalism" (intrusions which hinder a system's operation), and "spoofing" (electronically impersonating a computer to appear like another in order to gain access). One of the most infamous individuals was Kevin Mitnick, who from the early 1980s until 1995 carried out various schemes to defraud and obtain property by false pretenses from a number of major corporations using unauthorized access to computers of software manufacturers and illegal copying of software from these sites (among many other things). Millions of dollars was thought to have been stolen (Smith, Grabosky and Urbas 2004, p. 182). There have been some serious consequences of hacking, including death of patients where medical records have been altered in hospitals (Smith 1998, p. 195).

One of the features of the history of hacking is the change in the motivations behind the commission of acts. Early hackers in the 1960s "had a genuine belief in the liberating power of technology" based on principles such as freedom of information and unrestricted access to technology (Furnell 2010, p. 174). Over the last 50 years, however, the motivation for gaining access to computer systems without authorization has changed from liberation through more malicious motives involving the disruption of government activities to the financially oriented crimes of the twenty-first century. This period has also seen an exponential increase in attempts to identify users with greater certainty, including various forays into biometric user authentication systems (Smith 2007).

The proliferation of electronic funds transfer systems has enhanced the risk that such transactions may be intercepted and diverted. Between June and October 1994, a group of Russians attempted to steal approximately US$10.7 million from various Citibank customers' accounts in the United States by manipulating its computerized funds transfer system. One offender had been working in a Russian firm and gained access over 40 times to Citibank's funds transfer system using a personal computer and stolen passwords and account identification numbers. Using a computer terminal in his employer's office in St. Petersburg, he authorized transfers of funds from Citibank's head office in New Jersey to accounts which he and his co-conspirators held in Israel, California, Germany, the Netherlands, Switzerland, and Finland (Smith, Grabosky and Urbas 2004, p. 179).

Electronic fraud can also be directed at government agencies. In one case in Australia, a financial consultant formerly contracted to the Australian Government Department of Finance and Administration (as it then was known), was convicted on 25 September 2001 of defrauding the Australian government by transferring A$8.7 million electronically to private companies in which he held an interest. He did this by logging in to the department's computer network using another person's name and password. He was also able to obscure an audit trail through the use of other employees' login codes and passwords. He was sentenced to seven years and six months' imprisonment with a non-parole period of three years and six months (R. v Muir, Supreme Court of the Australian Capital Territory, 25 September 2001).

The advent of online banking now makes funds transfer fraud of considerable interest to criminals, as entire bank accounts are available to those who are able to compromise banking security and user authentication systems. Reliance on key-logging malware that can capture login and password combinations is the main risk for online banking that financial institutions are seeking to overcome through a variety of measures, including contactless cards and touch-screen computers that do not rely on keystrokes.

One of the technological developments within the financial services industry most apparent to the public has been the introduction of automated teller machines (ATMs) in the 1960s and, subsequently, electronic funds transfer at point of sale (EFTPOS) devices. Soon after they were introduced, an individual in Adelaide in South Australia closed his bank account but did not return his ATM card. He then used it offline to withdraw money from the ATM, owing to a design flaw in early machines which meant that when they were offline (unconnected to the bank) the machine was incapable of determining whether the cardholder had any account which remained current. He was convicted of larceny but on appeal argued that the bank had programmed the ATM to allow the withdrawal to take place and accordingly had agreed to the withdrawal. The court held that this did not amount to consent to the transaction and decided that he had acted fraudulently with intent permanently to deprive the bank of the funds (Kennison v Daire, High Court of Australia, 20 February 1986, 160 CLR 129).

More recently, the problem of plastic card skimming has arisen, in which devices with radio transmitters are attached to ATMs to trick unsuspecting users into inserting their cards into false facades placed on machines. A disguised micro-camera then records their PIN, and the offender can use the transmitted card details copied from the reader along with the PIN to create a counterfeit card that can then be used at another ATM to withdraw funds. A related, non-electronic crime problem concerns simple theft of funds from ATMs after the money has been dispensed. Both are examples of how enterprising criminals can make use of a new technology to steal money.

In one case, for example, a Malaysian national who had entered Australia on a tourist visa admitted to stealing A$623,426.91 from 315 Commonwealth, St. George, and Westpac Bank customers after having skimmed their cards at 36 ATMs around Sydney between May 2001 and November 2002. He then made 64 transactions over 18 months, depositing about A$200,000 in bank accounts held by the Hong Kong and Shanghai Banking Corporation in sums of less than A$10,000 to prevent the transactions being detected by Australia's Financial Intelligence Unit and Regulator. He was sentenced to three years' imprisonment at the New South Wales District Court on 30 October 2003 (Barker 2003).

Networking risks

Soon after the Internet was created, it was found that malicious code could be created that would affect its operation, sometimes with drastic consequences. The motivations for such conduct were varied and extended from simple nuisance-making, through curiosity and fame-seeking, to attempts at financially driven extortion. Various types of malware have been developed, including viruses, which use contaminated program elements to impede operation; worms, which attach themselves to other programs; bombs, which lay dormant until a predetermined time or event and then become active (such as the end of the millennium on 1 January 2000); and Trojan horse software, which is concealed within another program and when activated impedes the operation of the main program. The latest manifestation is spyware that scans computers for personal information that can then be used for the commission of other crimes (Furnell 2010, p. 185).

Telecommunications systems can also be used for harassing, threatening, or intrusive communications, from the traditional obscene telephone call to its contemporary manifestation of "cyber stalking," in which persistent messages are sent to an unwilling recipient. The first prosecution in England for cyber stalking occurred in the case of a computer programmer who sent his former girlfriend threatening emails after she ended their two-year relationship (Smith, Grabosky and Urbas 2004, p. 185). He was convicted of harassment by email under the Protection from Harassment Act 1997 (Eng.). Cyber stalking represents a clear case of technology-enabled crime, as it simply permits conventional stalking and harassment to be carried out more efficiently using the Internet.

Automated global risks

In the 1980s, as malware developed and as email usage became more widespread, cybercriminals began disseminating bulk email through the use of worms or Trojan horses with a view to incapacitating servers of target organizations. Motivations were often based on revenge or extortion, although some were politically driven attacks. An early case in the United States involved a former systems administrator for the United States District Court in Alaska, who launched three denial of service attacks against the United States District Court for the Eastern District of New York. He managed to overwhelm the court's server with emails to prove that it was vulnerable to external attack (Smith, Grabosky and Urbas 2004, p. 197).

The most recent problem concerns so-called "bot armies," in which thousands of remotely controlled computers (zombies) are made to bombard another computer or service, causing it to be overloaded and degrade or fail. Bot programs are codes or programs that operate automatically as agents for a user or another program. The first bot program was probably Eggdrop, which originated as a useful feature of Internet relay chat (IRC) in the early 1990s. Early bot programs were designed to allow IRC operators to script automated responses to IRC activities. As IRC gained popularity among Internet users, inappropriate behavior started to become a problem. Misbehaving users were ejected from IRC channels. As payback, some ejected users developed ways to attack the IRC channel, which led to the IRC wars that caused the first denial of service attacks in the mid-1990s (Choo 2007, p. 1). Membership of botnets ranges from hundreds to hundreds of thousands of zombies. Recently observed trends, however, suggest that botnet sizes have been decreasing so as to evade detection, since smaller botnets are often more difficult to detect. The level of sophistication of bot malware has, however, increased considerably in recent times.

Perhaps the most prolific crime risk associated with automated computing is identity crime. Identity crime facilitated through ICT has now become one of the most costly crime problems in developed societies, worth billions of dollars each year and in excess of 5 percent of the population being victimized (see Smith 2010a for a review). The Internet has provided an effective way in which personal information can be obtained and trafficked. In the first half of 2007, Symantec (2007) reported that the primary cause of data breaches that could facilitate identity theft was the theft or loss of a computer or other medium on which data were stored or transmitted such as a USB key or a backup device. It was found that the loss of such devices comprised 46 percent of all data breaches reported. Those seeking personal information online are also beginning to make use of an ever-expanding digital underground economy.

A recent form of online identity fraud is phishing, which involves the use of spam emails asking online banking customers and others to click on a hyperlink in the message to confirm their passwords and to provide other personal information. If they do so, they are diverted to the offender's website where the personal information is captured for later use in credit card and other fraudulent activities.

One of the first cases of phishing (then known as mirroring) to emerge in New South Wales in 2002 concerned offenders who created a phishing website, www.sydneyopera.org, that was used to divert customers away from the genuine site, www.sydneyoperahouse.com. The phishing site, however, had its own credit card booking arrangements, so that customers' money would be credited to the offenders' account. The offenders also created 23 sites mirroring opera houses in Europe, including in Paris and Vienna. The computer crime unit of the New South Wales Police Commercial Crime Agency contacted the FBI after tracing the bogus site to a Miami Internet server. It then moved to Santa Clara in California (Sydney Morning Herald 2002).

There has been a substantial increase in the number of phishing cases since this time. An industry group, the Anti-Phishing Working Group, has been collecting data on phishing since December 2003. In December 2003, 113 unique phishing attacks were reported, with eBay and Citibank being the two most targeted brands (Anti-Phishing Working Group 2003). By December 2013, this had increased to 52,489 unique phishing attacks being reported with 42,890 unique websites detected. In December 2013, the payment services sector remains the most targeted industry, followed closely by the financial services sector. The most recent development in phishing has been styled "spear-phishing and whale-phishing" in which individuals inside corporations, or of high net worth, are specifically targeted with attempts to gain access to corporate online banking systems, corporate VPN networks, and other online resources (Anti-Phishing Working Group 2013). In the case of phishing, opportunities have grown largely due to the absence of capable guardians, as phishing is difficult to detect and prosecution on such a large scale impossible to achieve.

The most recent forms of cybercrime rely on mobile and wireless technologies. Originally developed for cellular telephones, wireless connections to the Internet for use with laptop computers and PDAs have created new risks in an environment in which guardianship and security is less capable. Wireless networks create a number of vulnerabilities. Key among these is the fact that networks and their data can be accessed without physical access being required. This facility assists both the user and the criminal. The advent of wireless networking increases the likelihood of such information and associated tools being uploaded and downloaded. Enhanced methods of exploiting wireless vulnerabilities include drive-by subversion of wireless home routers through unauthorized access by mobile Wi-Fi clients and other forms of attacks. These include eavesdropping and sniffing of the VoIP calls, man-in-the-middle attacks, denial of service attacks, call interruption, and making "free" calls on VoIP networks built over wireless local area networks (Choo and Smith 2008). Of course, these latest risks remind us of earlier attempts to scan electromagnetic impulses for financial gain.

Finally, developments in cloud computing have created new crime risks. Cloud computing involves a pool of virtualized computing resources that allow users to gain access to applications and data in a web-based environment on demand (Choo 2010). Specific risks include unauthorized access to data held in external storage environments, virus infection and hosting of botnets, use of cloud services to launch brute force attacks (a strategy used to break encrypted data by trying all possible decryption key or password combinations) on various types of passwords, and concerns of data leakage. Already some of these risks have eventuated (Hutchings, Smith and James 2013).

Applying opportunity theory in cyberspace

This brief review of both technological developments in computing and communications and the criminal activities that they have spawned has shown clear indications of how opportunity theory can apply in practice. As new technological innovations have occurred, individuals have sought to compromise them for personal or financial gain. The motivations underlying some of these crimes have been constant, having their roots in greed, sexual desire, or socio-political change. Other innovations have altered people's motivations with, for example, early instances of unauthorized access to computers being driven by curiosity or a desire to enhance one's status, while more recent attempts at gaining access to information may be driven by financial gain, corporate espionage, or religious jihad. The industry and law enforcement response to cybercrime has often been slow and only achieved real results some time after the cybercrime problem became well established. Only rarely have new cybercrime risks been identified in advance of hardware or software being developed and implemented with success to prevent new forms of criminality from occurring.

Table 2.1 presents a diagrammatic illustration of how a selection of cybercrime types have adapted over time to accommodate new technologies and new opportunities for offending. It is possible to identify three phases applicable to the types of illegality in question with adaptations being driven largely by the need to secure new opportunities or to avoid detection by law enforcement.

The traditional response to cybercrime, as with other forms of crime, has been to devise a range of environmental measures that seek to make the commission of crime less attractive to potential offenders. Such measures operate on three levels (Clarke 1995). They seek to increase the levels of effort and skill required to commit offenses, to create a greater risk of apprehension of offenders, and to decrease the potential rewards

Table 2.1   Phases in the adaptation of cybercrime attacks

Cybercrime type	Phase 1	Phase 2	Phase 3
Interception	Postal, landline	EMR scanning	RFID cards
Phreaking	Black box	PABX	VOIP, Skype
Consumer scams	Door-to-door	Telemarketing	Online, mobile
Funds transfer fraud	Bank transfers	Payroll, invoicing	Online banking
Malware	Experimental	Disruption/extortion	Terrorism
ATM attacks	Robbery	Contact skimming	Remote attacks
Phishing	Simple – trading	Government targets	Extortion DDoS
Identity crime	Personal crime	Banking and finance	Government
Cyber terrorism	Intelligence	Target investigation	Mobile detonation
Cloud computing	Illegal data access	Data manipulation	Extortion

that offenders seek to derive from their illegal activities. If, however, it is assumed that potential offenders act on the basis of some rational calculation in which they balance up the likely risks and benefits to be derived from a potential course of conduct, then as some types of crime are seen to become too difficult to commit, other easier targets may be considered. Smith, Wolanin, and Worthington (2003) applied principles of crime prevention through environmental design to cybercrime risks and identified a number of ways in which crime displacement might occur. The ways in which displacement occurs also affect the time at which new or adapted forms of cybercriminality take place, for if crime prevention has been effective, it may take some considerable time for new patterns of offending to emerge.

Table 2.2 presents a range of strategies to minimize the risk of the occurrence of the ten cybercrime types examined in Table 2.1. These are considered in respect of Cohen and Felson's (1979) three patterns of routine activity relating to the presence of motivated offenders, the availability of suitable targets, and the absence of capable guardians against a violation. Had knowledge of the past been understood and acted upon, this could have provided more effective crime reduction approaches than have existed during the history of cybercrime.

In blocking cybercrime opportunities, for example, data security and user authentication controls represent the most appropriate ways in which crime can be prevented. These often entail technological solutions that are developed in direct response to the onset of new crime typologies. The challenge is to devise such technological approaches before the widespread onset of new cybercrime types, using knowledge of what has been effective, and, indeed, ineffective in the past. Strategies

Table 2.2 Crime reduction employing knowledge of the past

Cybercrime type	Opportunities	Motivations	Guardianship
Interception	RFID screening	Open government	Early notification
Phreaking	Detection/blocking	Low cost/free calls	Identity checks
Consumer scams	Spam blocking	Full employment	EFT monitoring
Funds transfer fraud	Password control	Staff satisfaction	User verification
Malware	Firewalls/filters	Refusing ransoms	Data monitoring
ATM attacks	Target hardening	Early detection	ATM security
Phishing	Risk awareness	Early detection	Email scanning
Identity crime	ID security	Full employment	Online verification
Cyber terrorism	Precursor controls	Anti-radicalization	Target surveillance
Cloud computing	Access control	Early detection	Data monitoring

that target offender motivations should seek to address the principal reasons for acting illegally, which, in recent decades, have invariably been economic in nature. Providing low-cost or free digital services and products can be an effective way in which to address economic crime motivations, while early detection should enable new crime types to be addressed promptly. Finally, surveillance and monitoring, which are greatly enhanced through the use of digital technologies, provide an effective deterrent to committing cybercrime, as long as the data surveillance that is undertaken is adequately monitored by those who can take action in appropriate circumstances.

The future trajectory of cybercrime

Over the ensuing years, a number of developments will occur that are likely to facilitate the commission of cybercrime (Choo, Smith and McCusker 2007). These include the integration of technology into everyday personal and business life, with people working from home and using the same systems for both business and private use, and increasing use of information and communications technologies by governments that require people to conduct transactions securely with government departments.

A major factor in technology-enabled crime threats remains the human element. There is likely to be a movement from syntactic (attacking the computer) to semantic (attacking the computer user) attacks. An example of this is the vast increase in phishing attacks that are essentially semantic attacks in which users of computers are targeted for the information they hold.

Globalization and the emergence of new economics, such as those in China and India where regulation and security may not be as well developed as in the West, will lead to new risks. Increasingly, the widespread use of broadband services will make the commission of some types of online crime easier and faster to commit. Because computers are often online 24 hours a day, and are used to download large files, risks of unauthorized access and threats from viruses increase. Similarly, increasing use of mobile and wireless technologies create new risks where these devices and technologies are insufficiently secured and protected from misuse. Because new hardware devices are portable, they are more easily stolen and may often not have passwords used or data encrypted. In 2005, one survey found that 22 percent of people reported losing their mobile devices, and, of those, 81 percent had not encrypted the information in any way (Millman 2005).

Increasing use of electronic payment systems and online banking as well as the development of new online payment systems such as e-gold and online gaming currencies will enable new forms of financial crime and money laundering to occur. In connection with this, changes in the ways in which people identify themselves when conducting online activities such as electronic banking, including the use of biometrics, will also provide avenues for misuse and compromise of new security measures.

Future risk areas

What then are the key areas of risk for the future? In relation to computer-facilitated fraud, attacks are moving away from large operations such as global spam attacks to smaller, more focused attacks on particular individuals who receive personally addressed invitations. Phishing will move to Voice over Internet Protocol (Vishing) and mobile SMSs (SMiShing) where phishing filters are less effective. Context-aware phishing will also emerge with images and content more easily able to trick users.

In relation to unauthorized access to networks, offenders will seek to gain access to computers to thwart physical security systems such as burglar alarms. Unauthorized access may be facilitated by corrupt insiders, including IT personnel who could create code to allow access at some future time, and displacement risks of violence may develop in which offenders seek to obtain access codes using threats of physical harm.

More sophisticated malware will be developed that can avoid detection by filters and can circumvent anti-malware products. Vulnerabilities will also arise from user-generated web content. Botnets will continue to constitute one of the Internet's biggest threats and will move to instant messaging and other peer-to-peer networks for command and control of infected systems. Ransomware attacks using encryption will also occur in commercial contexts. Industrial espionage and intellectual property infringement will pose threats for business, particularly with respect to hacking into unencrypted commercial-in-confidence emails and wireless communications, the use of insecure outsourcing, and theft of electronic trademarks and patents.

In relation to particular demographics at risk of harm, it is likely that cyber bullying based on text messaging and blogs will continue to evolve as will theft of increasingly smaller devices with access to unprotected data and personal information. Theft of virtual property from social networking sites such as World of Warcraft and Second Life will also take place (Warren and Palmer 2010). For example, police in China have investigated cases of virtual theft, including instances of organized gangs engaging in online robbery. In 2006, officers from Shenzhen arrested more than 40 suspects who were accused of stealing up to 700,000 yuan worth of virtual items from users of the social networking site "QQ" (Johnson 2007).

Offensive content will continue to be a problem, particularly for law enforcement, as cryptography is used by offenders to prevent access to child exploitation images. Live streaming of images of child abuse in chat rooms will also expand as will the use of ICT to organize racially motivated attacks. Organized crime groups will continue to be attracted by financially motivated scams and the use of denial of service attacks by botnets for commercial extortion. Money laundering using stored-value cards and mobile payments systems will also be used, and the Internet will continue to provide information for use by terrorist groups seeking to plan and execute attacks. In Australia, for example, an individual was convicted of plotting to bomb Australia's national electricity grid in October 2003 and sentenced to 20 years in prison for acts including downloading electricity grid maps from the Internet (2006 NSWSC, p. 691). Other threats will arise in connection with national information infrastructure, including disruption of supply chain management systems, financial sector networks, and power grids. The future may also see private sector companies being used as a launching pad for attacks.

This chapter has provided countless examples of how those with criminal motivations when provided with access to new technologies can adapt and expand their offending capabilities. If the chances of being detected are low, offending will become all the more likely. Unfortunately, the history of cybercrime has seen a number of examples of the ICT industry providing individuals with exactly the right kinds of tools to commit increasingly sophisticated and damaging acts. Had the pressures of commercial marketing and return on investment not been so strong, some of these criminal opportunities could have been averted. Changing the criminal motivations of people is difficult, even using the deterrent effects of criminal prosecution and punishment. Similarly, ensuring that cybercrime is prevented through adequate regulation and policing is often impractical, especially in the cross-border environment in which these crimes take place. Arguably, the most effective course of action is to design technologies that are unable to be used for criminal purposes, or if they are, that permit rapid discovery and rectification of the problem. This requires an understanding of the ways in which technologies have been misused in the past. I hope the present chapter will provide an abundance of instances for future designers to ponder when planning their next digital device.

Bibliography

Anti-Phishing Working Group (2003) *Phishing Activity Trends Report 4th quarter 2003*, New York: APWG http://www.antiphishing.org (accessed 13 January 2015).

Anti-Phishing Working Group (2010) *Phishing Activity Trends Report 4th quarter 2009*, New York: APWG http://www.antiphishing.org (accessed 13 January 2015).

Anti-Phishing Working Group (2013) *Phishing Activity Trends Report 4th quarter 2013*, New York: APWG http://www.antiphishing.org (accessed 13 January 2015).

Anti-Phishing Working Group (2014) *Phishing Activity Trends Report 1st quarter 2014*, New York: APWG http://www.antiphishing.org (accessed 13 January 2015).

Barker, G. (2003) "Counterfeit Credit Card Gangs on Rise," *Sunday Age*, 17 August: 7.

Choo, Kim-Kwang Raymond (2007) "Zombies and Botnets," *Trends and Issues in Crime and Criminal Justice* 333, Canberra: Australian Institute of Criminology.

Choo, Kim-Kwang Raymond (2010) "Cloud Computing: Challenges and Future Directions," *Trends and Issues in Crime and Criminal Justice* 400, Canberra: Australian Institute of Criminology.

Choo, Kim-Kwang Raymond and Smith, Russell G. (2008) "Criminal Exploitation of Online Systems by Organised Crime Groups," *Asian Criminology* 3: 37–59.

Choo, Kim-Kwang Raymond, Smith, Russell G. and McCusker, Rob (2007) "Future Directions in Technology-enabled Crime: 2007–09," *Research and Public Policy Series* 78, Canberra: Australian Institute of Criminology.

Clarke, Ronald V. (1995) "Situational Crime Prevention," in *Building a Safer Society: Strategic Approaches to Crime Prevention*, M. Tonry and D. P. Farrington (eds.), Chicago: University of Chicago Press, 91–150.

Clough, Brian and Mungo, P. (1992) *Approaching Zero: Data Crime and the Computer Underworld*, London: Faber and Faber.

Cohen, Lawrence E. and Felson, Marcus (1979) "Social Change and Crime Rate Trends: A Routine Activity Approach," *American Sociological Review* 44: 588–608.

Curran, James (2010) "Reinterpreting Internet History," in *Handbook of Internet Crime*, Y. Jewkes and M. Yar (eds.), Cullompton: Willan Publishing, 273–301.

Denning, Dorothy (2010) "Terror's Web: How the Internet is Transforming Terrorism," in *Handbook of Internet Crime*, Y. Jewkes and M. Yar (eds.), Cullompton: Willan Publishing, 194–213.

Felson, Marcus and Clarke, Ronald V. (1998) *Opportunity Makes the Thief: Practical Theory for Crime Prevention*, London: Home Office, Policing and Reducing Crime Unit.

Fitzgerald, Patrick and Leopold, Mark (1987) *Stranger on the Line: The Secret History of Phone Tapping*, London: The Bodley Head Limited.

Furnell, Steven (2010) "Hackers, Viruses and Malicious Software," in *Handbook of Internet Crime*, Y. Jewkes and M. Yar (eds.), Cullompton: Willan Publishing, 173–93.

Gibson, William (1984) *Neuromancer*, London: HarperCollins.

Hutchings, Alice, Smith, Russell G. and James, Lachlan (2013) "Cloud Computing for Small Business: Criminal and Security Threats and Prevention Measures," in *Trends and Issues in Crime and Criminal Justice* 456, Canberra: Australian Institute of Criminology.

Grabosky, Peter N. and Smith, Russell G. (1998) *Crime in the Digital Age: Controlling Telecommunications and Cyberspace Illegalities*, Sydney: Federation Press.

Johnson, B. (2007) "Virtual Robber Nabbed for Real," *Age (Melbourne)*, 16 November, http://www.theage.com.au/news/world/virtual-robber-nabbed-for -real/2007/11/15/1194766867217.html (accessed 13 January 2015).

Millman, R. (2005) "IT Managers Fail to Protect Mobile Devices," *SC Magazine*, 11 November.

O'Toole, G. J. A. (1978) *The Private Sector: Private Spies, Rent-a-cops, and the Police-industrial Complex*, New York: W. W. Norton and Company Inc.

Santayana, G. (1905) *The Life of Reason; or the Phases of Human Progress* vol. 1, New York: Dover Publications (1980 edn.).

Smith, Russell G. (1996) "Stealing Telecommunications Services," in *Trends and Issues in Crime and Criminal Justice* 54, Canberra: Australian Institute of Criminology.

Smith, Russell G. (1998) "The Regulation of Telemedicine" in *Health Care, Crime and Regulatory Control*, R. G. Smith (ed.), Sydney: Hawkins Press, 190–203.

Smith, Russell G. (2007) "Biometric Solutions to Identity-related Cybercrime," in *Crime Online*, Y. Jewkes (ed.), Cullompton: Willan Publishing, 44–59.

Smith, Russell G. (2010a) "Identity Theft and Fraud," in *Handbook of Internet Crime*, Y. Jewkes and M. Yar (eds.), Cullompton: Willan Publishing, 273–301.

Smith, Russell G. (2010b) "The Development of Cybercrime: An Opportunity Theory Approach," in *Crime Over Time*, R. Lincoln and S. Robinson (eds.), Newcastle upon Tyne: Cambridge Scholars Publishing, 211–36.

Smith, Russell G., Grabosky, Peter N. and Urbas, Gregor F. (2004) *Cyber Criminals on Trial*, Cambridge: Cambridge University Press.

Smith, Russell G., Wolanin, Nicholas and Worthington, Glenn (2003) "e-Crime Solutions and Crime Displacement," in *Trends and Issues in Crime and Criminal Justice* 243, Canberra: Australian Institute of Criminology.

Sydney Morning Herald (2002) "A Scam to Bring the House Down," *Sydney Morning Herald*, 28 August, http://www.smh.com.au/cgi-bin/common/printArticle .pl?path=/articles/2002/08/27/1030053059530.html (accessed 13 January 2015).

Symantec (2007) *Internet Security Threat Report*, Cupertino: Symantec.

Vaidhynathan, S. (2003) *Copyrights and Copywrongs: The Rise of Intellectual Property and How It Threatens Creativity*, New York: New York University Press.

Wall, David S. (2010) "Criminalising Cyberspace: The Rise of the Internet as a 'Crime Problem'," in *Handbook of Internet Crime*, Y. Jewkes and M. Yar (eds.), Cullompton: Willan Publishing, 83–103.

Warren, Ian and Palmer, Darren (2010) "Crime Risks of Three-dimensional Virtual Environments," in *Trends and Issues in Crime and Criminal Justice* 388, Canberra: Australian Institute of Criminology.

3
Ethical, Legal, and Methodological Considerations in Cyber Research: Conducting a Cyber Ethnography of www.bullying.org

Barry E. Cartwright

Introduction

As Lee (2013) points out in *Facebook Nation*, we live in an age of "total information awareness." With Google Street View, MySpace, YouTube, Facebook, Twitter, WikiLeaks, and recent revelations by Eric Snowden regarding the extent of the NSA's surveillance activities, it may seem whimsical to talk about privacy rights, informed consent, and other ethical concerns when it comes to conducting research in cyberspace. Nevertheless, the debate concerning ethics in cyber research continues unabated, as it affects most social researchers whose activities are governed by institutional research ethics boards and by the standards of their professional associations (Graber and Graber 2013; McKee and Porter 2009).

Mixed methods in cyber ethnography

Cyber communities, like real-world communities, have shared norms and values, ongoing (sustained) communication, a sense of commonality and belonging, and an agreed-upon "netiquette" (Hine 2000). They also establish community boundaries, enforce community norms, and exclude outsiders or others who fail to conform. Thus, it is said that these cyber communities can be studied using traditional ethnographic methods – for example, observing the setting and the social actors, recording and interpreting the narratives and histories of the actors, and analyzing how norms, values, and community boundaries are established and understood (Jankowski 2006; Mcmillan and Morrison 2006). The data set used in this cyber ethnography of www.bullying.org consisted of the first 475 messages posted on the website (as of 31 December 2002) and the last 475 messages posted on the site (as of 30 August 2006). This procedure captured two extended snapshots in time, providing an exact picture of what the messages on www.bullying.org looked like at the end of 2002 and what the messages on a more recent portion of the site looked like toward the end of 2006. The 950 messages were coded and categorized in NVivo (a computer software package designed to facilitate qualitative research) and then analyzed further in the SPSS (the Statistical Package for the Social Sciences). This approach could be described as a mixed or multiple methods study (Tashakkori and Teddlie 1998) or as complementary data analysis (Sudweeks and Simoff 1999).

The messages downloaded from www.bullying.org ranged in size from a few words to over ten pages in length, with some being little more than brief commentary and others rising almost to the level of life histories. None followed a preset format – the authors of messages could say as much or as little as they wanted. In many cases, information regarding age, gender, location, and reasons for posting the message were provided by the author. In almost as many cases, this information could only be inferred from a careful reading and rereading of the messages, if it could be inferred at all. Most messages were read eight or more times from beginning to end, often side by side, for purposes of comparison, identification of significant patterns or themes, verification of the identity of author, and assessment for identity play.

The messages were subjected to content analysis because they had elements or attributes that were physically present and countable, and it was possible to summarize the entire set of messages numerically (Berger 2000; Bryman, Teevan and Bell 2009; Neuendorf 2002). The messages were subjected to discourse analysis and narrative analysis, as a number of them were fully developed stories or narratives, with beginnings, middles and ends, often recounting experiences that occurred many years ago or that took place over many years (cf. Silverman 2003; Titscher et al. 2000).

The sample included messages from bullies, victims, bully-victims, bystander-observers, school teachers and school principals, police and probation officers, parents of bullies and victims, other relatives of bullies and victims, plus teachers, parents, and other relatives who were formerly bullies, victims, or bystanders themselves. Three hundred and sixteen messages were posted by adults describing experiences from long ago (in some cases, more than 40 or 50 years ago), offering fresh insight into how little the quality or quantity of bullying has actually changed throughout the years. The messages came from all ten provinces and three territories in Canada, 22 states in the USA, ten countries in Western and Eastern Europe, seven countries in Africa, five countries in Central and South America, four countries in Asia or South East Asia, and Australia and New Zealand. Apart from illustrating the vast potential of cyber research (Kraut et al. 2004; Joinson 2005), the diversity of these messages attests to cross-cultural and transnational concern with the issue of bullying.

Copyright law and cyber ethnography

The issue of copyright was raised by the university library in response to an innocuous question about how to format and acknowledge text messages taken word for word from www.bullying.org. The library insisted that these messages could not be used "as is" without express written consent of the site owners/operators, and that in the absence of such consent, the research findings would likely be unpublishable due to copyright infringement. To buttress their position, the university librarians pointed to the copyright symbol on the home page of the site and asserted that they were relying upon a previously obtained legal opinion from a Canadian copyright lawyer.

To preclude the possibility of copyright infringement, advice was sought from Michael Geist, who holds the Canada Chair in Internet and E-Commerce Law at the University of Ottawa's Law School. A written legal opinion was also obtained from Ann Carlsen, a Canadian lawyer who specializes in trademark and intellectual property law. According to the written legal opinion, the copyright symbol did not imply ownership of the messages, or for that matter, ownership of anything, unless copyright in fact existed. Site owners commonly display a copyright symbol or notice on their home page, but this usually indicates that copyright exists for the site itself. It does not imply that the site owners hold the copyright to the personal messages or stories that are posted on their site (Carlsen 2006). While some website owners might attempt to assert ownership rights to user content, the messages posted on sites such as www.bullying.org still belong to the individuals who posted them (Heilferty 2011; Moreno, Fost and Christakis 2008).

While the copyright still belonged to the author of each posting, it would have been impossible to seek individual permission from each author. There were literally hundreds of individuals in various countries around the world, most of whom had posted their messages anony-mously. Even if the means existed to make individual contact with the authors of the messages, the nature of the Internet is such that it would have been equally impossible to verify the identity of those who claimed to be the authors of the messages and thus ensure that they had the authority to grant permission to use their messages.

The issue, then, was how to make substantial use of the messages posted on www.bullying.org without engaging in copyright infringe-ment. The Canadian *Copyright Act* (RSC 1985, C-42) defines copyright infringement as doing anything that only the owner has the right to do, without the owner's consent (Government of Canada 1985). One of the sole rights that the owner has is to reproduce their work or any substan-tial part of their work. When it comes to what constitutes a "substan-tial part" of the work in question, Canadian courts have placed more emphasis on the quality than on the quantity of what was taken from the original work. Courts have also considered whether the copyright holder was adversely affected by the copying, whether the person copy-ing the material was doing so simply to save personal time and effort, and whether the work was being used in a similar fashion to that of the copyright owner (Carlsen 2006).

In circumstances where only a part of a message or story was being quoted, it is doubtful that there would be a substantial taking of the work, as long as its use did not adversely affect the copyright owner. While the messages on www.bullying.org were subject to copyright, the authors voluntarily placed these postings in a public venue, often intending to share them with others (cf. Heilferty 2011; Moreno, Fost and Christakis 2008; Martin 2012). Moreover, messages harvested from the site were not copied in order to save time and effort but rather so that they could be used in a totally different fashion from that of the copy-right owners. This research study was something that the owners of the copyright would be unable or unlikely to do themselves and something that researchers would be unable to do properly without copying the postings (Carlsen 2006).

Even if substantial parts of these postings were used to give "voice" to the research subjects, this would not necessarily imply that copyright infringement had taken place (Carlsen 2006; Geist 2006b; Geist 2006a). The Canadian *Copyright Act* (RSC 1985, C-42) states that "fair dealing for review or criticism does not infringe copyright" if the source of the work is acknowledged. The Supreme Court of Canada has stated that copying an entire work might still meet the definition of fair dealing for research purposes. If the researcher was making extensive use of text messages, but was putting them within the context of his or her own original thoughts, then this would be consistent with the principles of fair dealing, as long as the author and source of the downloaded messages were acknowledged (McKee and Porter 2009; Craig 2005).

As the researcher resided in Canada, the research took place at a Canadian university, and www.bullying.org was a Canadian-based website, it was decided that Canadian copyright law would take precedence if copyright infringement was claimed. Moreover, 94 percent of the messages posted on www.bullying.org came either from Canada, or else from the United States, the United Kingdom, Australia, or New Zealand, all of which have "fair dealing" provisions similar to those in Canada (McKee and Porter 2009; Geist 2013).

Ethical conduct and the requirement for informed consent

Informed consent is a central ethical principle in research scenarios that pose potential risk, harm, discomfort, or embarrassment to the research subjects (Thomas 1996; Comstock 2012). Internet research has drawn particular attention because of the ease with which "personal" messages can be downloaded, subjected to analysis, and quoted at length, without informing the research subjects that the study is taking place. It has been argued that the type of Internet research undertaken in this study of www.bullying.org could abrogate the right of the research subjects to know about the nature and duration of the research project, the potential risks and benefits, and what measures were being taken to ensure confidentiality (King 1996; Mann and Stewart 2000; Paechter 2012; Sharf 1999; Sharkey et al. 2011; Waskul and Douglass 1996).

It would have been impossible to obtain informed consent in this study because there were no means by which to verify with certainty the age or identity of the individuals giving such consent, short of conducting hundreds of offline interviews in some 29 countries around the world. According to information provided by the site participants, 284 were children when they posted their messages, while another 273 were teenagers. If informed consent was indeed required, then in many cases it would have been necessary to obtain the consent of the parents as well.

In any event, informed consent is not normally required when conducting research on public behavior in readily accessible public settings, where there is limited expectation of privacy and little if any risk of harm or embarrassment to the participants (Kitchin 2002a; Lindlof and Taylor 2002). Bullying.org is a readily accessible website where anyone can read or post a message without having to register or obtain a password. One of the first websites to appear in an Internet search using the keyword "bullying," the operators describe it as "the number one bullying site in the world." It is linked to a variety of similar sites, and its operators actively seek publicity around the world. Moreover, the site operators had already anonymized messages that were not submitted anonymously.

In the final analysis, this study of www.bullying.org was deemed to fall under the minimal risk provisions of the guidelines for *Ethical Conduct for Research Involving Humans* (Canadian Institutes of Health Research, Natural Science and Engineering Research Council of Canada and Social Sciences and Humanities Research Council of Canada 2007). The Research Ethics Board (REB) at the Canadian university where the study was conducted confirmed that it was excluded from the requirement for an ethics review and thus from the requirement for informed consent. The study used readily accessible archival materials posted in a public arena, it involved no interaction with the research subjects, and it posed no greater risk than what might normally be encountered by the research subjects in their daily lives (Moreno, Fost and Christakis 2008; Kitchin 2002a).

Some observers have argued that REB approval does not necessarily mean that the research is ethically or morally acceptable (Comstock 2012; Zimmer 2012). The Internet is an ever-changing landscape, and Board members might not be familiar with the techniques that cyber researchers employ to collect their data or the type of privacy settings that the researchers might be circumventing (McKee and Porter 2009; Palys and Atchison 2012). There is no question that the www.bullying. org community would rise to the definition of a "sensitive" or "at-risk" population (cf. Paechter 2012; Sharkey et al. 2011), in that 284 of the participants in the sample were children, 228 of the messages described bullying that involved the commission of one or more recognizable criminal offenses, and 94 messages talked about suicidal thoughts, self-harm, attempted suicide, and completed suicides. If the names, contact details, and other identifying features had not already been removed by site operators, the REB disposition in this case might well have been negative (Tilley and Woodthorpe 2011).

Privacy rights in the public domain

Some ethicists argue that website participants such as those on www.bullying.org have a "perceived" expectation of privacy (Martin 2012; King 1996; Zimmer 2012). The precise concepts of privacy and expectation of privacy can be difficult to pin down because interpretations vary significantly from individual to individual and from culture to culture – what some might regard as a highly personal secret, others might be willing to share publicly in cyberspace (Capurro 2002; Pomerance 2005). Indeed, millions of people seem quite willing to share personal information like their age, gender, photos, and contact details on social networking sites such as Facebook and MySpace (Raacke and Bonds-Raacke 2008). People can also be surprisingly forthcoming when it comes to the content of their home pages or web messages, freely sharing intimate details about their marital status, family problems, and medical concerns (Kennedy 2006; Lee 2005). Even those who do not have a home page or use social networking sites may voluntarily provide all sorts of revealing information about themselves in order to download free software or to gain access to a website that requires registration and a password (Klang 2004; Woo 2006).

While acknowledging that information posted on publicly accessible websites is in reality open to micro-observation and later inspection, a number of observers insist that researchers should disguise the source of the data in such a manner that members of a virtual community who subsequently recognized themselves or their own words in an article or website posting would feel that the location and identity of the group had been protected (King 1996; Waskul and Douglass 1996; cf. Boehlefeld 1996; Waruszynski 2002). However, many cyber ethnographers have felt comfortable in identifying the source of their data. Examples of studies where the source was identified include Kitchen's (2002b) study of an Alcoholics Anonymous site called Alt.Recovery.AA and Sanders' (2005) study of how sex-trade workers communicate with prospective clients through a website known as Punternet. In the case of www.bullying.org, any attempt to obfuscate the identity of the website would almost certainly have met with failure due to the site's prominence and public visibility.

The participants in www.bullying.org were clearly aware that their messages were being read. Forty-eight thanked the site or the other site participants for reading their messages, 54 specifically said they wanted their story told, 126 asked for a response or advice, 28 attempted to provide their complete contact details, and 10 mentioned that their stories had been published elsewhere. Of the 950 messages in the sample, 201 of them were in fact replies to earlier messages, often expressing sympathy, providing advice, or offering friendship and support. Notwithstanding arguments in favor of respecting personal privacy and obtaining informed consent, it is now widely accepted by courts and ethics review bodies in most countries that individuals posting messages in these public venues are often seeking public visibility, or if not, are at least aware of the public nature of their behavior (Heilferty 2011; Moreno, Fost and Christakis 2008; Martin 2012).

Discussion

There has been an ongoing (sometimes heated) debate about research ethics in cyberspace, but there has yet to be a definitive conclusion regarding what is acceptable and what is not. Some researchers have insisted that Internet users have a right to privacy – that messages posted on websites should be off-limits to social researchers unless informed consent is properly obtained and all members of the website have been notified that the research is taking place. Others have argued that public postings are by definition public – that they are no different from writing a letter to the editor of a newspaper or calling into a radio talk show.

Either the deontological (rule-based) framework or the teleological (harm vs. benefit-based) framework can be applied when addressing ethical concerns of this nature, regardless of whether the research is taking place in cyberspace or in a more conventional social domain (Thomas 1996; Palys and Atchison 2012). This cyber ethnography of www.bullying .org met the requirements of the deontological framework – it fell under the exemptions set forth in the Canadian guidelines for *Ethical Conduct for Research Involving Humans* and received written confirmation from the university's REB that it was excluded from the requirement for an ethics review and the requirement for informed consent. As noted previously, it also complied with the requirements for fair dealing for research purposes under the Canadian *Copyright Act* (RSC 1985, C-42).

Nevertheless, there may be a genuine perception of privacy in smaller websites with closed membership and especially those that have registration requirements and password protection (Heilferty 2011). Cyber researchers would be well advised to consider the situational context, carefully examining such issues as restricted site access, perceptions of privacy held by the site participants, and whether they are conducting research on a vulnerable or "at-risk" population (Heilferty 2011; Zimmer 2012; Nissenbaum 2011). It should be remembered that the site participants still have "moral rights" – including the right to privacy – even though these rights may not be enshrined in law in the country or countries where the website and the cyber researcher are domiciled (McKee and Porter 2009; Comstock 2012). In other words, researchers should try to avoid unnecessary intrusion into "private" spaces, and if they find such intrusion necessary, to limit the disclosure of intimate or highly sensitive information (Hull, Lipford and Latulipe 2011). This cyber ethnography of www.bullying.org also met the requirements of the teleological framework. There was minimal risk of harm or embarrassment to the research subjects, and the likelihood of their being identified or contacted as a result of the study was virtually nonexistent (Heilferty 2011; Lee 1993). The study contributes to the understanding of bullying and victimization by bringing forward the voices, stories, impressions, and advice of those who have had personal, firsthand experience with bullying. In particular, it provides new insights that challenge the oft-expressed notion that bullying is on the upswing, or that current manifestations of bullying are more serious than earlier ones (cf. Haber and Glatzer 2007; Marsh et al. 2004; Olweus 1993). It also offers an explanation for why anti-bullying programs have routinely produced negligible to modest results at best (Rigby 2002; Twemlow and Sacco 2008). Moreover, it is hoped that this work will serve as a guidepost or template for other cyber researchers, bearing in mind that research decisions need to be made on a case-by-case basis (Sharkey et al. 2011) due to the complex nature of cyberspace and its many diverse cyber communities.

Bibliography

Berger, A. A. (2000) *Media and Communication Research Methods: An Introduction to Qualitative and Quantitative Approaches,* Thousand Oaks, CA: Sage Publications.

Boehlefeld, S. P. (1996) "Doing the Right Thing: Ethical Cyberspace Research, *The Information Society* 12(2): 141–52.

Bryman, A., Teevan, J. J. and Bell, E. (2009) *Social Research Methods* 2nd Canadian edn, Toronto: Oxford University Press.

Canadian Institutes of Health Research, Natural Science and Engineering Research Council of Canada and Social Sciences and Humanities Research Council of Canada (2007) *Tri-Council Policy Statement: Ethical Conduct for Research Involving Humans,* http://www.pre.ethics.gc.ca.proxy.lib.sfu.ca/english/pdf/TCPS October 2005_E.pdf (accessed 7 June 2008).

Capurro, R. (2002) "Privacy: An Intercultural Perspective," *Ethics and Information Technology* 7(1): 37–47.

Carlsen, A. (2006) *Legal Opinion: The Application of the Copyright Act to Thesis Content,* provided to author 29 August 2006.

Comstock, G. (2012) *Research Ethics: A Philosophical Guide to the Responsible Conduct of Research,* Cambridge, UK: Cambridge University Press.

Craig, C. (2005) "The Changing Face of Fair Dealing in Canadian Copyright Law: A Proposal for Legislative Reform," in *In the Public Interest: The Future of Canadian Copyright Law,* M. Geist (ed.), Toronto: Irwin Law, 437–61.

Geist, M. (2013) "Fairness Found: How Canada Quietly Shifted from Fair Dealing to Fair Use," in *The Copyright Pentalogy: How the Supreme Court of Canada Shook the Foundations of Canadian Copyright Law*, M. Geist (ed.), Ottawa: University of Ottawa Press, 157–86.

Geist, M. (2006a) "After Access: Canadian Education and Copyright Reform," *Education Canada* 46(3): 22–5.

Geist, M. (2006b) *Personal Communication re: Canadian Copyright Law*, provided to author 6 August 2006.

Government of Canada (1985) *Copyright Act* (RSC 1985, C-42).

Graber, M. A. and Graber, A. (2013) "Internet-Based Crowdsourcing and Research Ethics: The Case for IRB Review," *Journal of Medical Ethics* 39(2): 115–18.

Haber, J. and Glatzer, J. (2007) *Protect Your Child from Teasing, Taunting and Bullying for Good*, New York: Penguin Books.

Heilferty, C. M. (2011) "Ethical Considerations in the Study of Online Illness Narratives: A Qualitative Review," *Journal of Advanced Nursing* 67(5): 945–53.

Hine, C. (2000) *Virtual Ethnography*, London: Sage.

Hull, G., Lipford, H. R. and Latulipe, C. (2011) "Contextual Gaps: Privacy Issues on Facebook," *Ethics and Information Technology* 13(4): 289–302.

Jankowski, N. W. (2006) "Creating Community with Media: History, Theories and Scientific Investigations," in *Handbook of New Media: Social Shaping and Social Consequences of ICTs*, L. A. Lievrouw and S. M. Livingstone (eds.), updated student edn, London: Sage, 55–64.

Joinson, A. N. (2005) "Internet Behaviour and the Design of Virtual Methods," in *Virtual Methods: Issues in Social Research on the Internet*, C. Hine (ed.), Oxford, UK: Berg, 21–34.

Kennedy, H. (2006) "Beyond Anonymity, or Future Directions for Internet Identity Research," *New Media & Society* 8(6): 859–76.

King, S. A. (1996) "Researching Internet Communities: Proposed Ethical Guidelines for the Reporting of Results," *The Information Society* 12(2): 119–28.

Kitchin, H. A. (2002a) "The Tri-Council on Cyberspace: Insights, Oversights, and Extrapolations," in *Walking the Tightrope: Ethical Issues for Qualitative Researchers*, W. C. Van den Hoonaard (ed.), Toronto: University of Toronto Press, 160–73.

Kitchin, H. A. (2002b) "Alcoholics Anonymous Discourse and Members' Resistance in a Virtual Community: Exploring Tensions Between Theory and Practice," *Contemporary Drug Problems* 29(4): 749–77.

Klang, M. (2004) "Spyware: The Ethics of Covert Software," *Ethics and Information Technology* 6(3): 193–202.

Kraut, R., Olson, J., Banaji, M., Bruckman, A., Cohen, J. and Couper, M. (2004) "Psychological Research Online: Report of Board of Scientific Affairs," Advisory Group on the Conduct of Research on the Internet, *American Psychologist* 59(2): 105–17.

Lee, H. (2005) "Implosion, Virtuality, and Interaction in an Internet Discussion Group," *Information, Communication & Society* 8(1): 47–63.

Lee, N. (2013) *Facbook Nation: Total Information Awareness*, New York: Springer.

Lee, R. M. (1993) *Doing Research on Sensitive Topics*, London: Sage Publications.

Lindlof, T. R. and Taylor, B. C. (2002) *Qualitative Communication Research Methods* 2nd edn, Thousand Oaks, CA: Sage Publications.

Mann, C. and Stewart, F. (2000) *Internet Communication and Qualitative Research: A Handbook for Researching Online,* Thousand Oaks, CA: Sage Publications.

Marsh, H. W., Parada, R. H., Craven, R. G. and Finger, L. (2004) "In the Looking Glass: A Reciprocal Effect Model Elucidating the Complex Nature of Bullying, Psychological Determinants, and the Central Role of Self-concept," in *Bullying: Implications for the Classroom,* C. E. Sanders and G. D. Phye (eds.), San Diego, CA: Elsevier/Academic Press, 63–106.

Martin, K. E. (2012) "Diminished or Just Different? A Factorial Vignette Study of Privacy as a Social Contract," *Journal of Business Ethics* 111(4): 519–39.

McKee, H. A. and Porter, J. E. (2009) *The Ethics of Internet Research: A Rhetorical, Case-Based Process,* New York: Peter Lang.

Mcmillan, S. J. and Morrison, M. (2006) "Coming of Age with the Internet: A Qualitative Exploration of How the Internet Has Become an Integral Part of Young People's Lives," *New Media & Society* 8(1): 73–95.

Moreno, M. A., Fost, N. C. and Christakis, D. A. (2008) "Research Ethics in the MySpace Era," *Pediatrics* 121(1): 157–60.

Neuendorf, K. A. (2002) *The Content Analysis Guidebook,* Thousand Oaks: Sage Publications.

Nissenbaum, H. (2011) "A Contextual Approach to Privacy Online," *Daedalus* 140(4): 32–48.

Olweus, D. (1993) *Bullying at School: What We Know and What We Can Do,* Oxford, UK: Blackwell.

Paechter, C. (2012) "Researching Sensitive Issues Online: Implications of a Hybrid Insider/Outsider Position in an Ethnographic Study," *Qualitative Research* 13(1): 71–86.

Palys, T. and Atchison, C. (2012) "Qualitative Research in the Digital Era: Obstacles and Opportunities," *International Journal of Qualitative Methods* 11(4): 352–67.

Pomerance, R. M. (2005) "Redefining Privacy in the Face of New Technologies: Data Mining and the Threat to the 'Inviolate Personality'," *Canadian Criminal Law Review* 9(3): 273–94.

Raacke, J. and Bonds-Raacke, J. (2008) "MySpace and Facebook: Applying the Uses and Gratifications Theory to Exploring Friend-networking Sites," *Cyberpsychology & Behavior* 11(2): 169–74.

Rigby, K. (2002) *A Meta-evaluation of Methods and Approaches to Reducing Bullying in Pre-schools and Early Primary School in Australia,* Canberra: Commonwealth Attorney-General's Department.

Sanders, T. (2005) "Researching the Online Sex Work Community," in *Virtual Methods: Issues in Social Research on the Internet,* C. Hine (ed.), Oxford, UK: Berg, 67–79.

Sharf, B. F. (1999) "Beyond Netiquette: The Ethics of Doing Naturalistic Discourse Research on the Internet," in *Doing Internet Research: Critical Issues and Methods for Examining the Net,* S. Jones (ed.), Thousand Oaks, CA: Sage Publications, 243–56.

Sharkey, S., Jones, R. A., Smithson, J., Hewis, E., Emmens, T., Ford, T. and Owens, C. (2011) "Ethical Practice in Internet Research Involving Vulnerable People: Lessons from a Self-harm Discussion Forum Study (SharpTalk)," *Journal of Medical Ethics* 37(12): 752–58.

Silverman, D. (2003) "Analyzing Talk and Text," in *Collecting and Interpreting Qualitative Materials* 2nd edn, N. K. Denzin and Y. S. Lincoln (eds.), Thousand Oaks, CA: Sage Publications, 340–63.

Sudweeks, F. and Simoff, S. J. (1999) "Complementary Explorative Data Analysis: The Reconciliation of Quantitative and Qualitative Principles," in *Doing Internet Research: Critical Issues and Methods for Examining the Net*, S. Jones (ed.), Thousand Oaks, CA: Sage Publications, 29–55.

Tashakkori, A. and Teddlie, C. (1998) *Mixed Methodology: Combining Qualitative and Quantitative Approaches*, Thousand Oaks, CA: Sage Publications.

Thomas, J. (1996) "Introduction: A Debate about the Ethics of Fair Practices for Collecting Social Science Data in Cyberspace," *The Information Society* 12(2): 107–118.

Tilley, L. and Woodthorpe, K. (2011) "Is It the End of Anonymity as We Know It? A Critical Examination of the Ethical Principle of Anonymity in the Context of 21st Century Demands on the Qualitative Researcher," *Qualitative Research* 11(12): 197–212.

Titscher, S., Meyer, M., Wodak, R. and Vetter, E. (2000) *Methods of Text and Discourse Analysis*, London: Sage Publications.

Twemlow, S. W. and Sacco, F. C. (2008) *Why School Antibullying Programs Don't Work*, Lanham, MD: Jason Aronson.

Waruszynski, B. T. (2002) "Pace of Technological Change: Battling Ethical Issues in Qualitative Research," in *Walking the Tightrope: Ethical Issues for Qualitative Researchers*, W. C. Van den Hoonaard (ed.), Toronto: University of Toronto Press, 152–9.

Waskul, D. and Douglass, M. (1996) "Considering the Electronic Participant: Some Polemical Observations on the Ethics of On-Line Research," *The Information Society* 12(2): 129–40.

Woo, J. (2006) "The Right Not to Be Identified: Privacy and Anonymity in the Interactive Media Environment," *New Media & Society* 8(6): 949–67.

Zimmer, M. (2012) "'But the Data Is Already Public': On the Ethics of Research in Facebook," *Ethics and Information Technology* 12(2): 313–25.

4

That Cyber Routine, That Cyber Victimization: Profiling Victims of Cybercrime

Fernando Miró Llinares

Introduction: New crimes, new victims, in a new space – Cyberspace 2.0

Beyond the discussion of the most appropriate term to refer to crimes perpetrated on the Internet, and even the question of the different defining characteristics for each type of crime (Furnell 2002; Clough 2010; Wall 2007), today it is generally accepted that the term "cybercrime" also encompasses a set of crimes that affect basic personal interests such as privacy, dignity, honor, liberty, and sexual liberty (Clough 2010) within the framework of cyberspace use as a new environment for social intercommunication. Some of these that we can call "social cybercrimes" are cyber bullying (Smith et al. 2008), cyber stalking (Basu and Jones 2007), online child grooming (Berson 2003; McAlinde 2006), and different types of cyber harassment (Henson 2011). Although these types of crimes may be understood as transpositions of crimes committed in physical space, it is universally accepted among scholars that such crimes display multiple characteristics related to the profile of the offender, the victim, and the way the crime is expressed that are different from physical crimes because they are perpetrated in "a new space." The reason for this is obvious but bears repeating: the Internet is a new space for communication that is different from physical space, and this fact changes the way in which social events are expressed as well as the crime committed within this new space.

This fact may explain why some of the most widely used criminological approaches to analyzing cybercrime are crime theories, particularly Routine Activity Theory (RAT). Since Grabosky (2001) suggested using them to understand cybercrime, many papers have taken advantage of this focus to try to understand to what extent the environment, in this case the Internet, influences whether a crime occurs or affects the crime that does occur. This study also uses this approach, but its objective is not to demonstrate RAT's explanatory power for cybercrime but rather to find risk factors for victimization by harassment on the Internet that are related to users' everyday use of information and communication technologies. This will lead us, following analysis of how other authors have applied this theory to cybercrime, to propose a reconceptualization for cybercrime of the characteristics which, according to the theory's traditional description of predatory crimes in physical space, make a target suitable.

Victimization by social cybercrime and routine activities in cyberspace: From VIVA to ISI

The literature analyzing victimization by social cybercrime, especially in the case of minors, and more particularly in the case of victimization by cyber bullying, generally focuses on analysis of demographic, personality, and social relationship risk factors for victimization (Peter and Valkenburg 2006; Ybarra and Mitchell 2007; Mitchell, Wolak and Finkelhor 2008; Ybarra, Espelage and Mitchell 2007; Hinduja and Patchin 2008). In recent years, other studies have tried to relate victimization by social cybercrimes to other factors linked to daily activities of their potential victims as has been done in relation to other forms of cyber victimization that affect users' property (Choi 2008; Bossler and Holt 2009; Pratt, Holtfreter and Reisig 2010; Yucedal 2010; Ngo and Paternoster 2011). The first of these was Marcum's (2008) study, followed by Holt and Bossler (2009), Reyns (2010), and Ngo and Paternoster (2011).

Although victimization behaviors and the definitions of RAT's independent variables are different in each of the studies, they have elements in common, particularly in how they operationalize the theory, that should be highlighted. The first of these is the fact that all of these authors attempt to define variables relevant to each of RAT's constructs as defined by Cohen and Felson (1979): the motivated offender, target suitability, and absence of a capable guardian. This is logical insofar as the goal is to validate RAT's usefulness for explaining this type of cybercrime, but we should bear in mind that the methodology used by all of these studies consists of asking users who have or have not been victimized about their daily usage, so that it will always be easier to obtain information about the target's characteristics than about the offender's or the capable guardian's. In fact, another common characteristic of the studies applying RAT to victimization by social cybercrimes is that they identify as characteristics of the motivated offender, or the capable guardian, variables that we may also identify as characteristics of a suitable target. This is the case when Marcum (2008) defines the variable "exposure to the motivated offender," as hours spent on the Internet using particular communication tools such as electronic mail, instant messaging, and so on. Similarly, Ngo and Paternoster (2011), and also Holt and Bossler (2009), posit that users are exposed to motivated offenders when they engage in interactive activities such as shopping, using social networks, electronic mail, playing games, and so on, but also when they use a high-speed Internet connection, engage in deviant behavior on the Internet, and have a low level of Internet skills. Finally, Reyns (2010) distinguishes between "exposure to the motivated offender" and "proximity to the motivated offender," describing the first case as time spent on the Internet and engaging in activities such as use of social networks, and the second as adding strangers to social networks and using certain websites to search for friends.

In our opinion, and as explained below, interaction and communication with others is a way of behavior that refers to visibility of users in cyberspace. Following Yucedal (2010), the mere act of accessing cyberspace means exposure to the motivated offender (p. 43), just as it does in the physical world. The relevant aspect of exposure to the motivated offender is determined by visibility (Cohen et al. 1981, p. 507), defined as the offender knowing of the existence of the potential target.

Something similar occurs with the variables referred to as "capable guardianship," identified with the subject's own actions to defend himself against possible attacks. When Holt and Bossler (2009) speak of a capable guardian, they define it as the use of security software such as anti-virus programs or firewalls; Ngo and Paternoster (2011) propose the use of security software, but they add the subject's own skills as well as having received education about Internet risks; and Reyns (2010), though with different actions, also suggests behaviors that the subject himself might undertake to protect itself, such as maintaining privacy on social networks and using an online profile tracker to view who has visited his social network. In those cases, more than the action of potential guardianships, the authors are measuring actions of self-protection. Different is the case of Marcum (2008), who measures the vigilance that third parties may exercise over victims and their goods.

With the above in mind, we do not argue that it is impossible to undertake an analysis of all the elements of the RAT crime triangle, but another possible use of RAT that is functionally oriented toward prevention is to focus on determining the suitable target's characteristics in cyberspace. If the end goal is to understand which of the victim's routine activities influence his victimization, it is appropriate to focus on the suitable target and its particular characteristic elements, while, as Yar (2005) has proposed, taking into account that the target is in a different environment.

Felson (1998) used the acronym VIVA to convey that in order for a target to be considered suitable, it must have *value* from the offender's perspective, *inertia*, physical *visibility*, and *accessibility*. As Yar (2005) points out, it is necessary to review the characteristics of suitable targets in cyberspace, while thinking that distance and time have a different structure on the Internet (Miró 2011). In this way, elements such as inertia lose their importance, and others, such as visibility and accessibility, should be specified and redefined based on the functioning of cyberspace. Furthermore, we exclude the element "value," although we know how important it is, because the main objective of our study is to identify which of the victim's routine activities may increase the possibility of victimization on the Internet.

The first element of the VIVA acronym that must be reconceptualized is accessibility. First of all, we might understand that cyberspace is a communication environment in which data must be introduced to be entered. Although we are in physical space and we "have things" in physical space, we "are" in cyberspace, and there we can have data that informs about our patrimony, our honor, our privacy. The decision to enter cyberspace and bring anything into it is sometimes not made on our own, other times it is completely voluntary, and often our own involuntary actions lead to the introduction of our image, data, and so on. But generally our own routine activities will determine what is introduced into cyberspace: having personal photos in the tablet, having real data in the computer, and the like. It is also true that some "goods" as patrimony or privacy can be harmed without any introduction of data in the Internet, but it is indubitable that if some elements of privacy, intimacy, or others are introduced in cyberspace and are accessible to potential offenders, it is easier to use them for a concrete attack.

Accessibility in cyberspace also depends on the way we protect our computer systems from the free access of others, especially using software security programs (anti-virus, anti-spyware, firewalls, etc.).

Although some authors have seen this as constituting a capable guardian (Holt and Bossler 2009; Marcum, Ricketts and Higgins 2010; Ngo and Paternoster 2010), these types of software are really elements that must be incorporated into the object itself in order to function, often by the user itself, and given that in Cohen and Felson's (1979) conception the guardian is someone other than the victim, we might view these as elements that, conceptually, define the suitability of an object and not its external protection, and thus they are really elements or characteristics of the suitable target. This is what we call "self-protection" to the extent that a target will be more suitable for an offender when it is less protected by anti-virus software but also when it engages in risky routines such as using pirated software or not updating the operating system.

When goods are introduced, voluntarily or not, and if they are not protected, they are exposed to risk, but they will only be suitable targets when they become visible to the offender. According to Yar (2005), cyberspace is a public medium, and therefore all targets are visible globally. But because everybody is on the Internet, the only users that are visible for potential offenders are users that interact with others, communicating with other users via the many tools available on the Internet (electronic mail, chat rooms, forums, instant messaging, social networks, etc.), using other services in cyberspace, such as online shopping, downloading files, watching videos, doing online banking, and so on. The more interaction there is, the greater the visibility, the greater the contact with multiple users, and therefore the more contact with potential offenders.

Thus, we shift from the acronym VIVA to the acronym ISI. The suitability of a potential target in cyberspace will depend on its having been *introduced* onto the Internet (which may be determined by the actions of the potential victim), whether its computer system is more or less *self-protected and is easy or difficult to access,* and the *interaction* of the user who makes itself visible to potential motivated offenders.

Social cyber victimization in a Spanish sample: A study

In view, then, of the scientific landscape and the need to determine victimization risk factors related to routine activities through the prism of the new realm of criminal opportunity in cyberspace, we conceived a study whose principal objectives would be to determine the prevalence of social cybercrime and to explain which routine activities in cyberspace increase the risk of victimization by these forms of cyber harassment. With these goals, and based on the theoretical and empirical background discussed above, we began with three starting hypotheses for the study. The first is linked to the introduction of goods into cyberspace; as argued above, the fact that cyberspace must be "entered" means that targets must be introduced, voluntarily or involuntarily, leading to the following hypothesis: the more targets are introduced into cyberspace, the greater the risk of victimization. The second hypothesis is based on the idea that the user, by interacting in cyberspace, defines its previously introduced targets' level of visibility; that is to say, it becomes more or less perceptible to offenders as a function of interaction, according to the following formula: the more interaction in cyberspace, the greater the risk of victimization. Finally, the third hypothesis is based on the idea that in cyberspace the victim can make its computer system less accessible against possible attacks using protective software, and so when there is less self-protection, the risk of victimization is greater.

Methodology

Sample

The study sample was made up of 500 Spanish adult participants, of whom 222 (44.4%) were men and 278 (55.6%) were women, with a median age of 40.21 (SD = 12.57). To select the subjects, we performed, following statistical norms, a probabilistic sampling stratified by sex, age, and autonomous region of residence. The criteria for participants' inclusion were that they: 1) use the Internet for a minimum of eight hours per week and 2) be between the ages of 18 and 65, for the reasons discussed above (INE 2011).

Variables

Dependent variable

The dependent variable "cyber victimization by online harassment," which is both categorical and dichotomous, was obtained through six answers to survey questions about victimization by online harassment. The experienced behaviors that constituted victimization are: having received repeated contact from someone after requesting that it stop; having been seriously threatened; having been threatened with revealing damaging information or causing some harm; having had personal information made public without consent; having one's image used or identity stolen; and having been insulted or falsely accused.

When respondents answered affirmatively to any of these behaviors, they were categorized in the dependent variable as "victims"; participants who did not experience any of these behaviors were categorized as "non-victims."

Independent variables

The independent variables included were created by the revision of RAT in cyberspace: "introduction," "interaction," and "self-protection."

"Introduction" refers to the goods that a person transfers, voluntarily or involuntarily, from physical to cyberspace. To measure it, we created a scale with the following nine items with dichotomous (yes/no) answers: having files on one's computer containing passwords, personal photos, intimate photos, personal videos, or sensitive business information; using real personal information to open accounts on social networks; providing real personal information through networks; providing real personal information through forums; and providing passwords. The score was obtained by adding the responses to each of the items (yes = 1, no = 0), with a maximum value of nine points.

The "interaction" variable was operationalized in two variables, distinguishing between activities related to communication in cyberspace and activities related to communication with strangers. Thus, "personal interaction" is defined as the actions of a person in cyberspace to communicate with other people that make him/her a visible user. We created a numerical value by adding users' responses to nine items, taking into account that the range of measurement is dichotomous (yes = 1, no = 0). The items were: use of electronic mail, chat rooms, instant messaging, social networks, forums, and videoconferencing; downloading files; consuming pornography; and playing online video games. The "interaction with strangers" variable refers to the actions taken by a user to contact strangers. This numeric variable was created by adding respondents' answers to five dichotomous items: using contact networks, contacting strangers through social networks, contacting strangers through instant messaging, opening or downloading links or files sent by strangers via email, and opening or downloading links or files sent by strangers via instant messaging.

Finally, for the "self-protection" variable, we chose to define it negatively in order to adapt it to the results of the third stated hypothesis. Thus, "no self-protection" means actions taken/not taken by an Internet user to protect his goods. Following the same method as in the preceding variables, this variable was created by adding the responses obtained for five dichotomous items: not having anti-virus software, using pirated software, using the same password for everything, not changing passwords at least annually, and having public accounts.

Instrument

For this study, an ad hoc survey was developed with input from professionals in various areas: jurists, criminologists, and methodologists. The survey was made up of four types of questions. A filter question asked, "How many hours per week do you spend connected to the Internet?" with the goal of including only subjects who spent at least eight hours on the Internet per week; following this were three questions about socio-demographic characteristics; then a series of questions about routine activities on the Internet; and finally, the survey included 14 questions about economic and social cyber victimization of which, for the present article, we will only consider those questions related to one form of social victimization: online harassment.

Procedure

To collect the information, we contracted the services of an outside company (Investgroup), which carried out the computer-assisted telephonic administration of the instrument using the CATI (Computer Assisted Telephone Interviewing) system, including both land lines and mobile telephones. The duration of each administration varied from eight to 15 minutes, and the survey was administered between 1 October and 15 October 2013.

Results

With respect to the dependent variable, the descriptive analyses showed, first, that 21 percent ($n = 105$) of the participants have experienced at least one of the specific forms of online harassment. Analyzing the data in detail, we can observe from Table 4.1 that "making information public without consent" behavior was the most often repeated, as we found that 10.2 percent ($n = 51$) of respondents reported having experienced it at least once. Following in order, "having received repeated contact from someone after requesting that it stop" was the next most-experienced behavior (10%, $n = 50$), followed by "having been seriously threatened" (3.2%, $n = 16$), "having one's image used or identity stolen" (3.2%, $n = 16$), "being falsely accused or insulted" (1.2%, $n = 6$), and "having felt intimidated" (1%, $n = 5$).

The descriptive analyses of the independent variables shows (Table 4.2) a certain variability when it comes to acts of introducing personal

Table 4.1 Prevalence of online harassment in the sample

Cyber-Victimization By Online Harassment	21% ($n = 105$)*
Making information public without consent	10.2% ($n = 51$)
Repeated unwanted contact	10% ($n = 50$)
Threats	3.2% ($n = 16$)
Identity theft	3.2% ($n = 16$)
Insults	1.2% ($n = 6$)
Intimidation	1% ($n = 5$)

*The percentage of cyber-victimization by online harassment corresponds to the percentage of subjects who have experienced at least one of the behaviours included in the category.

Table 4.2 Descriptive analyses of the independent variables

Independent Variables	n	χ	D.T.	S^2	Min	Max	Z of K-S (p)
Introduction	500	2.8	1.5	2.5	0	8	2.802 (0,000)
Personal interaction	500	3.8	1.7	3.0	0	9	2.562 (0,000)
Interaction with strangers	500	1.2	0.6	0.4	0	4	10.600 (0,000)
No self-protection	500	1.6	1.1	1.2	0	5	4.148 (0,000)

information into cyberspace, as participants' responses ranged between 0 and 8, with a median of 2.8 (D.T. = 1.5). There seems to be a greater tendency toward actions of communicating with others, where the median of the "personal interaction" variable was 3.8 (D.T. = 1.7). Nevertheless, none of the participants in the sample responded affirmatively to all of the behaviors included in the "interaction with strangers" variable, and the data show a low variability, with a median of 1.2 (D.T. = 0.6). Finally, the results of the "no self-protection" variable indicate that a segment of Internet users takes no action to protect its interests, as we obtained a median point value of 1.6 (D.T. = 1.1). With respect to the results of the K-S test to analyze the variables' normal distribution, they show that none of the variables is distributed normally, and for that reason we chose to use the non-parametric Mann-Whitney U-test for the bivariate analysis.

The results of the bivariate analysis (Table 4.3) indicate significant relationships between the dependent variable and three of the four independent variables analyzed. Specifically, victimized people are the ones who introduce more targets (UM–$W = 15803,00$; $p = 0,000$), engage in more personal interaction activities (UM–$W = 13973,00$; $p = 0,000$), and interact more with strangers (UM–W = 15655,00; p = 0,000) in comparison with participants categorized as non-victims, because the former are more visible in cyberspace than the latter. On the other hand, we did not find significant differences in what victims and non-victims did to safeguard their interests (UM–W = 18633,00; p = 0,097).

Finally, to determine the weight of the variables that affect victimization by online harassment and establish a predictive model of victimization, we performed a multiple logistic regression analysis. We chose to use a simultaneous method which does not prioritize the order of the variables so that we could analyze their influence all at once within the model.

The variables included in the model improve the null model significantly (χ^2 = 50.773; p = 0.000), and the Hosmer-Lemeshow χ^2 statistic shows that the model offers a good fit for the data (χ^2 = 7.692; p = 0.464). Finally, Nagelkerke's generalized R^2 coefficient of determination presents a value of 0.15 once the four predictive values are included in the model. Based on a threshold or cut-off probability of P(Y=1) = 0.5, the overall percentage of correct classification was 79.8 percent, and it improved the correct classification of victims and non-victims (79.8% and 97.7% respectively) (Table 4.4).

Table 4.3 Relationship between independent variables and the dependent variable

Independent Variables	U	p
Introduction	15803.00	0.000
Personal interaction	13973.00	0.000
Interaction with strangers	15655.00	0.000
No self-protection	18633.00	0.097

Table 4.4 Model

Model	
Omnibus	Step: χ^2 = 50.773 (p = 0.000)
	Block: χ^2 = 50.773 (p = 0.000)
	Model: χ^2 = 50.773 (p = 0.000)
–2LL	463.183
R2 Nagelkerke	0.150
Hosmer and Lemeshow	χ^2 = 7.692 (p = 0.464)
Classification	Global: 79.8%
	Non-victim: 97.7%
	Victim: 79.8%

Table 4.5 Variables included in the equation

| | B | E.T. | Wald | gl | Sig. | Exp(B) | C.I. 95% for EXP(B) | | Probability |
							Upper	Lower	
Introduction	0.961	0.333	8.331	1	0.004	2.613	1.361	5.017	72.32%
Personal interaction	0.548	0.224	5.967	1	0.015	1.730	1.114	2.685	63.3%
Interaction with strangers	0.202	0.099	4.165	1	0.041	1.224	1.008	1.486	55.04%
No self-protection	0.209	0.093	5.078	1	0.024	1.232	1.028	1.478	55.2%
Constant	-2.705	0.365	54.795	1	0.000	0.067			

All the variables ("introduction," "personal interaction," "interaction with strangers," and "no self-protection") predict the dependent variable ($p < 0.05$), so that subjects who introduce more targets into cyberspace become more visible through interaction with others (known or strangers), and protect themselves less have a greater probability of being victimized (Table 4.5). Taking into account the odds ratio, for each unit of increase in the "introduction" variable, a subject has a 72.3 percent (OR = 2.613) greater probability of online harassment victimization. Similarly, for each unit of increase in "personal interaction" and "interaction with strangers," the probability of victimization increases by 63.3 percent and 55.04 percent respectively (OR = 1,730 y OR = 1,224). Finally, for each unit of increase of "no self-protection," the subject has a 55.2 percent (OR = 1,232) probability of experiencing victimization.

Discussion

One out of every five users over the age of 18 has experienced at some point in his or her life some form of online harassment. Comparing our results with other studies, we can see that they are close to those found by other authors. Finn (2004) obtained values that varied between 10 percent and 15 percent. These slightly lower results might be due to the fact that they were obtained in 2004, when Internet use was not as extensive as it is now. A closer result can be found in the 18.9 percent obtained by Holt and Bossler (2009), the 20.1 percent obtained by Reyns (2010), the 20.6 percent obtained by Yucedal (2010), and the 7.6 percent to 20.9 percent obtained by Ngo and Paternoster (2011).

The similarity of results for social cybercrime victimization is diminished when compared with other studies' results for victimization by specific forms of harassment. In the case of threats, the percentages are significantly lower than those obtained by Bocij (2003), between 40 percent and 49 percent compared to 3.2 percent. Closer to our result, although still higher, is the 4.4 percent obtained by Reyns (2010). The same is true for insults, where Bocij (2003) obtained 24.4 percent and Ngo and Paternoster (2011) 7.6 percent, compared with 1.0 percent in the present study; for identity theft, 9 percent obtained by Bocij (2003) compared with 3.2 percent; and repeated unwanted contact, for which Reyns (2010) obtained 23.3 percent compared with 10 percent. The dynamic is repeated for cases of cyber sexual harassment, if we compare 2.2 percent with the 10–20 percent obtained by Marcum (2008), 13.9 percent by Reyns (2010), and 14.7–20.9 percent by Ngo and Paternoster (2011).

The disparity in these last results may have two reasons. The first and perhaps most significant may be methodological differences in the studies, specifically the means of measuring victimization. For this study, we chose particularly significant harassment behaviors, leaving aside other behaviors that might disturb or bother the victim but that we felt did not rise to the same level. It is clear, in any case, that all of this would be corrected if operative definitions and standard measuring tools existed that would allow for direct comparison of results at the national and international level. The second reason for this disparity may be related to the sample: most studies use samples of young university students, while ours is the adult population between 18 and 65. And even if risk is determined not by age but by the routine activities associated with age (Alshalan 2006), it is true that there will be greater victimization with greater use and exposure to cyberspace, which is more common among young people. In any case, the goal of our study was precisely to confirm this hypothesis, and we therefore believe that an analysis based on the entire adult population is preferable. It is evident, in any case, that more studies of this type are needed, particularly a similar study of the population between the ages of 13 and 18 which asks these young people about their routine activities in cyberspace.

Regarding the most important part of the study, the model, we can clearly conclude that the results confirm the initially stated hypothesis. With greater introduction of goods and interests into cyberspace, greater interaction with known and unknown users, and less self-protection, we find a greater probability of experiencing victimization. The "introduction" variable turns out to be the greatest predictor of online harassment victimization, raising the probability of victimization by 72.3 percent. Hence, more introduction leads to a greater probability of becoming a victim of some form of online harassment. This is consistent with the idea that if the target has not been introduced into cyberspace, it cannot be the object of an attack.

In looking at preventing victimization, it will be important to warn users of the risks involved in introducing personal goods into cyberspace. Providing personal information via the Internet's various personal communication tools creates risk, just as earlier studies (Marcum 2008; Marcum et al. 2010) had indicated, on the benefit of creating policies designed to change the current culture regarding sharing private information with third parties. Another risk that users may not be aware of is that when we connect our devices to the Internet, whether computer, tablet, or smart phone, all the information contained on them is available to hackers and other strangers in cyberspace. These risks may be avoided with acts as simple as saving information on external hard drives. Finally, for the use of personal information to open accounts on social networks or email accounts, in addition to the precautions that users may adopt, capable guardians or place managers must be integrated into the prevention process. After all, the user merely enters information solicited by the service provider, who may or may not make it public at a later point. Taking into account the relationship between this variable and victimization in future studies, we must consider more focus on these issues, changing service providers' policies by requiring them to adopt stricter security measures or require less personal data from users.

The second hypothesis, that more interaction leads to a greater probability of experiencing online harassment because it makes the user more visible to potential offenders, has also been proved. Consistent with the results obtained in earlier studies, interaction with strangers increases the risk of victimization by 55 percent (Marcum et al. 2010; Reyns 2010). Nevertheless, contact with strangers was conceptualized differently in those studies than it is here. Marcum et al. (2010) take the view that the act of contacting strangers is one more element of the suitable target, which is different from Reyns (2010), who identifies adding strangers to social networks and using webpages to make friends as an element of proximity to the motivated offender. In our opinion, contacting strangers is a behavior that makes targets more visible for others. And the same would be true for personal interaction; it increases the risk of victimization by 63 percent. These results also demonstrate that, in addition to the type of interaction that occurs in cyberspace, the quantity of different activities is also important, and therefore subjects who engage in more interaction activities have a greater probability of experiencing any of the online harassment behaviors evaluated here.

Finally, the third variable, self-protection, also turns out to be significant, in that adopting fewer protection measures leads to a greater risk of experiencing online harassment. In spite of the contradictions found in studies of the use of protection programs, according to the results of our study, not having anti-virus software and other non-protective behaviors turn out to be a risk factor, as found by Choi's study (2008) and in contrast to Holt and Bossler (2009). It is true that these programs are designed to stop malware attacks, but there are types of harassment attacks that require the use of these programs for an initial attack. We might think, for example, about the use of Trojan software to access computer systems in order to obtain information, such as private photos that can be used later to blackmail. The conclusions are clear for prevention strategies: it is necessary to educate users about the use of protection programs and to warn them of the risks of using pirated software, the need to close access to social network accounts, and the importance of good password management (not using the same password for everything, and changing them frequently). Although these recommendations have been made on other occasions, they have almost always been associated with the risk of victimization by hacking, or other cyber attacks, and it is important to make it clear that they also increase the risk of experiencing harassment.

Conclusions

This study supports the thesis that the victim is a key element in the production of the criminal event particularly on the Internet, where he determines his own risk zone by incorporating certain goods into cyberspace; interacting with others, particularly strangers; and not utilizing all possible means of self-protection. Based on this, the necessity emerges, beyond adopting the measures proposed above as Internet security policies, to devote resources to educating users about the risks inherent in certain activities. But above all, given the limitations of this study, what is most necessary is to continue researching these issues, at least in some of the following directions.

First, it is essential to continue periodic evaluations that allow us to better understand the phenomenon, especially as information and communications technologies are in a state of permanent evolution. To the extent that cybercrime is influenced by daily use of ICT, it will also continue to change, which is especially evident if we consider that the use of such technologies continues to expand.

Second, we must expand the sample to include subjects for whom cyberspace has been, and is, a fundamental element of their education and leisure time – digital natives (Prensky 2001). It would be advisable to evaluate how natural familiarity with ICT is related to processes of victimization, especially if we consider that despite their facility with navigating the Web from having grown up with the Internet, they have not been educated about technology use and do not know the essential *key* steps to preventing certain threats that, because distances do not exist in this new means of social intercommunication, may come from anywhere.

Bibliography

Alshalan, A. (2006) *Cyber-crime Fear and Victimization: An Analysis of a National Survey*, Starkville, MS: Mississippi State University.

Basu, S. and Jones, R. (2007) "Regulating Cyberstalking," *Journal of Information, Law and Technology* 2(1): 1–30.

Berson, I. R. (2003) "Grooming Cybervictims: The Psychosocial Effects of Online Exploitation for Youth," *Journal of School Violence* 2(10): 5–18.

Bocij, P. (2003) "Victims of Cyberstalking: An Exploratory Study of Harassment Perpetrated Via the Internet," *First Monday*, 8(10), http://www.firstmonday.org (accessed 12 February 2015).

Bossler, A. and Holt, T. (2009) "On-line Activities, Guardianship, and Malware Infection: An Examination of Routine Activities Theory," *International Journal of Cyber Criminology* 3(1): 400–20.

Choi, K. (2008) "Computer Crime Victimization and Integrated Theory: An Empirical Assessment," *International Journal of Cyber Criminology* 2(1): 308–33.

Clough, J. (2010) *Principles of Cybercrime*, Cambridge: Cambridge University Press.

Cohen, L. E., Kluegel, J. R. and Land, K. C. (1981) "Social Inequality and Predatory Criminal Victimization: An Exposition and Test of a Formal Theory," *American Sociological Review* 46(5): 505–24.

Cohen, L. and Felson, M. (1979) "Social Change and Crime Rate Trends: A Routine Activity Approach," *American Sociological Review* 44: 588–608.

Felson, M. (1998) *Crime and Everyday Life* 2nd edn, Thousand Oaks, CA: PineForge Press.

Finn, J. (2004) "A Survey of Online Harassment at a University Campus," *Journal of Interpersonal Violence* 19(4): 468–83.

Furnell, S. (2003) "Cybercrime: Vandalizing the Information Society," *LNCS* 2722: 8–16.

Grabosky, P. (2001) "Virtual Criminality: Old Wine in New Bottles?," *Social & Legal Studies* 10: 243–9.

Hinduja, S. and Patchin, J. (2008) "Cyberbullying: An Exploratory Analysis of Factors Related to Offending and Victimization," *Deviant Behavior* 29(2): 129–56.

Holt, T. and Bossler, A. (2009) "Examining the Applicability of Lifestyle-routine Activities Theory for Cybercrime Victimization," *Deviant Behavior* 30(1): 1–25.

Marcum, C. (2008) "Identifying Potential Factors of Adolescent Online Victimization for High School Seniors," *International Journal of Cyber Criminology* 2(2): 346–67.

Marcum, C., Ricketts, M. and Higgins, G. (2010) "Assessing Sex Experiences of Online Victimization: An Examination of Adolescent Online Behaviors Using Routine Activity Theory," *Criminal Justice Review* 35(4): 412–37.

McAlinden, A. M. (2006) "'Setting Em Up': Personal, Familial and Institutional Grooming in the Sexual Abuse of Children," *Social and Legal Studies* 15(3): 339–62.

Miró, F. (2012) *El Cibercrimen. Fenomenología y Criminología de la Delincuencia en el Ciberespacio*, Madrid: Marcial Pons.

Mitchell, K., Wolak, J. and Finkelhor D. (2008) "Are Blogs Putting Youth at Risk for Online Sexual Solicitation or Harassment?," *Child Abuse and Neglect* 32(2): 277–94.

Ngo, F. and Paternoster, R. (2011) "Cybercrime Victimization: An examination of Individual and Situational Level Factors," *International Journal of Cyber Criminology* 5(1): 773–93.

Peter, J. and Valkenburg, P. (2006) "Adolescents' Exposure to Sexually Explicit Material on the Internet," *Communication Research* 33(2): 178–204.

Pratt, T. C., Holtfreter, K. and Reisig, M. D. (2010) "Routine Online Activity and Internet Fraud Targeting: Extending the Generality of Routine Activity Theory," *Journal of Research in Crime and Delinquency* 47(3): 267–96.

Prensky, M. (2001) "Digital Natives, Digital Immigrants," *On the Horizon* 9(5): 1–6.

Reyns, B. (2010) *Being Pursued Online: Extent and Nature of Cyberstalking Victimization from a Lifestyle/Routine Activities Perspective*, doctoral dissertation submitted to the Graduate School of the University of Cincinnati.

Reyns, B., Henson, B. and Fisher, B. (2011) "Being Pursued Online: Applying Cyberlifestyle–Routine Activities Theory to Cyberstalking Victimization," *Criminal Justice and Behavior* 38(11): 1149–69.

Smith, P. K., Mahdavi, J., Carvalho, M., Fisher, S., Russell, S. and Tippett, N. (2008) "Cyberbullying: Its Nature and Impact in Secondary School Pupils," *Journal of Child Psychology and Psychiatry* 49(4): 376–85.

Wall, D. (2007) *Cybercrime: The Transformation of Crime in the Information Age*, Cambridge, UK: Polity Press.

Yar, M. (2005) "The Novelty of 'Cybercrime': An Assessment in Light of Routine Activity Theory," *European Journal of Criminology* 4(2): 407–27.

Ybarra, M., Espelage, D. L., and Mitchell, K. (2007) "The Co-Occurrence of Internet Harassment and Unwanted Sexual Solicitation Victimization and Perpetration: Associations with Psychosocial Indicators." *Journal of Adolescent Health* 41(6,Suppl): S31–S41.

Ybarra, M. and Mitchell, K. (2007) "Prevalence and Frequency of Internet Harassment Instigation: Implications for Adolescent Health," *Journal of Adolescent Health* 41(2): 189–95.

Yucedal, B. (2010) "Victimization in Cyberspace: An Applicaiton of Routine Activity and Lifestyle Exposure Theories." Electronic Thesis or Dissertation. Kent State University, 2010. https://etd.ohiolink.edu/.

Part II
Contemporary Cybercrime Risks

5
Organized Cybercrime and National Security

Peter Grabosky

Acknowledgment

An earlier version of this chapter appeared in Korean Institute of Criminology Research Report Series 13-B-01 *Information Society and Cybercrime: Challenges for Criminology and Criminal Justice*. Korean Institute of Criminology and International Society of Criminology, Seoul, 2013.

Introduction

While much cybercrime is committed by individuals acting alone, a significant amount is accomplished by offenders acting in concert. These groups have tended to vary significantly in terms of their structures, their goals, the criminal activities in which they engage, and their organizational life courses. The nature of these collectivities, and the question of whether organized cybercrime constitutes a national security threat, are the subjects of this chapter. The answer will depend on one's definitions of "national security" and "organized cybercrime." Each of these concepts is problematic; the meaning of national security has been stretched significantly in recent years, while conceptions of organized crime (terrestrial or in cyberspace) have been overly narrow. The chapter concludes that some forms of organized cybercrime can indeed threaten national security, both in a more conventional sense and in ways previously overlooked.

National security

"National security" is a term used loosely and for many purposes, not all of them legitimate. Traditionally, national security meant the capacity to deter or to resist the invasion of one's territorial borders by foreign military, naval, or air forces. Nowadays, the concept has expanded to embrace a range of factors that might support this capacity, including threats to public health, to education, to welfare, to social cohesion, and to the national economy. In recent years, senior US military officials have spoken of the national debt and global warming as significant national security threats.

One should be cautious about uncritically accepting pronouncements relating to security risks. Historically, national security has been invoked as a justification for domestic political repression, to chill democratic political debate, or to distract public and media attention from shortcomings in governance. The Nixon administration sought to terminate the investigation of the Watergate break-in and cover-up on national security grounds. Appeals to national security to justify military intervention are sometimes based on fabricated or exaggerated evidence of threat. The 2003 US invasion of Iraq was rationalized by the threat of weapons of mass destruction in the hands of Saddam Hussein (Bamford 2004). This alleged threat proved to have been greatly overstated. More recently, national security pretexts in the United States have served to obscure the contours of (and thereby inhibit debate on) the massive electronic surveillance program undertaken by the National Security Agency (*New York Times* 2013).

Organized crime

"Organized crime" is a term that means many things to many people. Klaus von Lampe's website (http://www.organized-crime.de/organized-crimedefinitions.htm) contains 150 definitions. Most if not all of these tend to be mired in the twentieth century, too narrow in scope, and too constrained by ideology. Most fail to capture the very significant changes in organizational form that have occurred since the beginning of the digital age.

Traditional definitions of organized crime tended to be based on the profit motive. However, even the most insightful observers of "terrestrial" organized crime note the intrinsic attraction of excitement, comradeship, and other non-material values to participants. Similarly, a great deal of organized criminal activity on the Internet is driven primarily by non-monetary considerations. These include the quest for intellectual challenge, individual or group notoriety, lust (in the case of organized paedophile activity), ideology, rebellion, and curiosity. Today, there are many criminal organizations active in cyberspace which do not exist to

enrich their members; nor do they practice violence or engage in bribery. Moreover, the traditional view of criminal organizations consisting
of full- or part-time professional criminals was also somewhat simplistic.
In cyberspace, no less than in terrestrial space, some criminal organizations have explicit or implicit membership but may also include a variety of hangers-on, camp followers, consultants, and accomplices, some
of whom will be well aware of their complicity in criminal enterprise,
while others may not.

Conventional organized crime was characterized as monolithic and
hierarchical in nature. The classic "mafia model" that served as the basis
of Cressey's (1972) analysis four decades ago conjured up visions of eth-
nically based, pyramidal organizations ruled by "godfathers" or "Mr.
Bigs." By the 1990s, observers of criminal organizations reported that
rather than the work of formal, enduring structures, a significant amount
of criminal activity was undertaken by loose coalitions of smaller groups
converging temporarily to exchange goods and services (Halstead 1998).
The idea of vertically integrated enterprises thus gave way to the metaphor of networks (Williams 2001), which provided a foundation for
contemporary thinking about the interrelationships *within* organized
criminal groups and *between* individual groups (Morselli 2009). Digital
technology also facilitates new organizational forms, such as short-term
opportunistic "swarms" which can serve criminal ends and which differ significantly from the traditional hierarchical paradigm of the mid-
twentieth century (Choo and Grabosky 2014).

Today, it requires an exceptionally closed mind to deny that states
are also capable of criminal acts. Throughout recorded history, crimes
by state actors have occurred in times of peace as well as during armed
conflict. In recent years, there have been allegations of drug manufacture and trafficking, illicit arms transfers, and counterfeiting by agents
of the Democratic People's Republic of Korea (Bradsher 2006; Perl 2007).
States have also engaged periodically in kidnapping and assassination,
at home and abroad. Today, the business of the International Criminal
Court is booming, notwithstanding the refusal of some states to submit
to its jurisdiction.

Just as the distinctions between public and private sectors have
blurred in recent years with regard to legitimate activities such as
public-private partnerships, contracting out, and other mechanisms for
the co-production of governance (Grabosky 1995), there is a long tradition of public/private collaboration in criminal activity to advance
state interests. Administrators of the French Concession in Shanghai
during the 1920s and 1930s relied upon the Green Gang to suppress

industrial unrest and regulate drug markets (Martin 1996; Wakeman 1995, pp. 123–30). The US Central Intelligence Agency engaged burglars and criminals around the world to conduct break-ins and kidnappings and enlisted mafia members in an unsuccessful attempt to assassinate Cuban President Fidel Castro (Weiner 2007, pp. 186, 199). Toward the end of the apartheid era, South African state security engaged criminal groups to assist with "sanctions busting" and with resisting ANC insurgents (Standing 2003).

Organized cybercrime, whether it is the work of "conventional" criminal offenders, nation-states, or some public/private hybrid, entails a diverse set of organizational forms and motives. The following paragraphs describe a number of organizations whose activities have been reported in recent years. The cases in question are not the product of a representative sample; rather, they have been selected to illustrate the diversity of organized cybercriminal exploits by state and non-state actors alike.

Wonderland was a members-only group that exchanged illicit images of children until its interdiction by a multinational police investigation named Operation Cathedral on 2 September 1998. Simultaneous raids in 14 countries revealed a group of 180 individuals from 49 countries around the world who collectively possessed over 750,000 illicit images of children and over 1,800 digitized videos depicting child abuse. The group had been established in the mid-1990s to facilitate file sharing of the images and videos (Russell 2008; Graham 2000). It was a very sophisticated collective that vetted prospective members, encrypted the information shared among them, and periodically rotated the physical location of the servers that supported the group's activities.

Anonymous is a loose group of anarchists based largely on a shared ethos of mischief and resentment of authority who engage in what Denning (2001) referred to as "hactivism." The participants' prevailing ethos of iconoclasm, if not nihilism, began to focus on prominent symbols. The chosen methods were website defacements and distributed denial of service attacks, complemented by online verbal abuse (Olson 2012). Not surprisingly, the website of the US Central Intelligence Agency represented an attractive target. Imbued with the hacker ethos that information should be free, the group also targeted the Church of Scientology because of its secrecy, the Motion Picture Association of America for its proprietary commercialism, and became a supporter of WikiLeaks. When the US government prevailed upon various electronic payment service providers to discontinue processing of contributions to WikiLeaks after its publication in 2010 of secret US State Department messages,

Anonymous orchestrated denial of service attacks against the complying sites (Coleman 2011).

Drink or Die was an international group of copyright pirates who illegally reproduced and distributed software, games, and movies over the Internet. They were motivated less by profit than by their desire for recognition as the first group to distribute a perfect copy of a newly pirated product. Founded in Moscow in 1993, the group expanded internationally within three years, with members in more than 12 countries, including Britain, Australia, Finland, Norway, Sweden, and the United States. Its approximately 65 members were technologically sophisticated and included IT professionals skilled in security, programming, and Internet communications, each performing specialized roles (US Department of Justice 2002; Lee 2002; McIllwain 2005; Urbas 2006).

The Ukrainian ZeuS Group, software engineers in Eastern Europe, had refined malware known as the ZeuS virus. This malicious code was used by one group of Ukrainian hackers to gain access to the computers of individuals employed in a variety of small businesses, municipalities, and non-government organizations in the United States. Target computers were compromised when the victim opened an apparently benign email message. With access to the victim's bank account numbers and password details, principals in the Ukraine were able to log on to the target organizations' bank accounts. Accomplices of the Ukrainian principals placed notices on Russian language websites inviting students resident in the United States to assist in transferring funds out of the country. These so-called "mules" were provided with counterfeit passports and were directed to open accounts in false names in various US financial institutions. When principals in the Ukraine transferred funds from legitimate account holders to the mules' accounts, the mules were instructed to move the funds to accounts offshore, or in some cases, to smuggle the funds physically out of the United States (FBI 2012).

Dark Market was a forum for the exchange of stolen credit card and banking details, malicious software, and related technology. Its website provided the infrastructure for an electronic bazaar – a meeting place for buyers and sellers of the illicit material. The forum was founded in May 2005 in order to take advantage of the criminal opportunities presented by the advent of electronic banking and the increasing use of credit and debit cards. Banking and card details were illicitly obtained by various means, including surreptitious recording with "skimming" devices, unauthorized access to personal or business information systems, or techniques of "social engineering" where victims were persuaded to part with the details at the request of an ostensibly legitimate source.

At its peak, Dark Market was the world's pre-eminent English language "carding" site, with over 2,500 members from a number of countries around the world, including the United Kingdom, Canada, the United States, Russia, Turkey, Germany, and France. Shortly thereafter, having been infiltrated by an undercover FBI agent, the market ceased operation (Glenny 2011).

Operation Olympic Games is reportedly a collaboration between the US National Security Agency and its Israeli counterpart, Unit 8200, intended to disrupt the Iranian nuclear enrichment program. It allegedly involved the clandestine insertion of an extremely complex and sophisticated set of software into communications and control systems at the Natanz nuclear facility. The software reportedly includes a capacity to monitor communications and processing activity, as well as the ability to corrupt control systems at the facility. The operation succeeded in delaying the progress of uranium enrichment through remote-controlled destruction of a number of centrifuges used in the process. The secrecy surrounding the operation was compromised in part when the malicious software escaped because of a programming error. Neither the United States nor the Israeli governments are inclined to discuss the operation (Sanger 2012; Zetter 2014).

GhostNet was the name given by a group of Canadian researchers in 2010 to a cyber espionage operation apparently operating from commercial Internet accounts in China. The hackers compromised government computers in over 100 countries on several continents; prominent among them was India. They also targeted emails from the server of the Dalai Lama (Markoff and Barboza 2010). The Chinese government denied involvement, and there was no conclusive evidence to the contrary. There was, however, some evidence of government complicity. Chinese officials have confronted expatriate dissidents returning to China with transcripts of Internet chats in which they were involved during their absence. Whether the activity in question was the work of patriotic hackers acting unilaterally, or skilled individuals with guidance from state authorities who were otherwise acting at arm's length, remains unclear. Canadian investigators claim to have found evidence of links to two individuals in the underground hacking community of the PRC (Information Warfare Monitor 2010).

PLA Unit 61398. In February 2013, the information security company Mandiant reported that a large-scale program of industrial espionage had been undertaken since 2006 by Unit 61398 of the People's Liberation Army (Mandiant Intelligence Center 2013; Sanger, Barboza and Perlroth 2013). Based in Shanghai, this organization is alleged to have acquired a massive volume of data from a wide variety of industries in English-speaking countries. Information alleged to have been taken includes technical specifications, negotiation strategies, pricing documents, and other proprietary data. One of the alleged targets, a major US beverage manufacturer, was planning in 2009 what was to have been the largest foreign purchase of a Chinese company to date. It was reported that an apparently innocuous email to an executive of the US company contained a link, which, when it was opened, allowed the attackers access to the company network. Sensitive information on pending negotiations was reportedly accessed by Chinese intruders on a regular basis; the purchase did not eventuate. It is unclear whether the unit is staffed exclusively by military personnel or includes civilian contractors. The US government filed criminal charges against five PLA officers, who were charged in absentia by the US Department of Justice in 2014 (Schmidt and Sanger 2014).

The Prism Program entails systematic harvesting of digital information by the US National Security Agency (NSA). Outlined five years earlier by Bamford (2008), further classified details of the scheme were disclosed in 2013 by whistleblower Edward Snowden (Gidda 2013). Part of a larger NSA program of data capture, storage, and analysis, Prism was assisted by some of the world's more prominent IT companies, including Microsoft, Google, Yahoo!, Facebook, Pal Talk, YouTube, Skype, AOL, and Apple. The program reportedly captured and stored a wide range of data, including email, chat-room exchanges, voice over Internet protocol (VoIP), photos, stored data, file transfers, video conferencing, and other social network content (Greenwald 2014). The precise nature of the engagement between the agency and the IT companies remains unclear. In some instances, industry cooperation was required by secret court order; some degree of voluntary cooperation may also have occurred. The legality of such data collection under US law has been challenged; it seems likely that the activity has breached the law of a number of nations whose citizens' communications were intercepted (Donohue 2013).

Public-private partnerships in cybercrime

One characteristic of state or state-sponsored cybercrime appears to be its hybrid organizational form. Try to envision a continuum of state-private interaction, from state ignorance of private criminal activity at one extreme to state monopoly of criminal activity at the other. In between these polar extremes, one might find state incapacity to control private illegality; the state turning a "blind eye" to the activity in question; tacit encouragement of non-state crime; active sponsorship by the state; loose cooperation between state authorities and private criminal actors; then formal collaboration between state and non-state entities.

One notes that Russian "patriotic hackers" were allegedly involved in cyber attacks on government servers in Estonia. It has also been suggested that Chinese civilian IT specialists working in one or more universities are linked with the People's Liberation Army. It has been reported that the North Korean Army has trained hackers to commit fraud for the purpose of raising revenue. The North Korean state was allegedly involved with a hacking ring that exploited online gaming sites and generated US$6 million in cash (Greitens 2012). The US National Security Agency spends millions of dollars engaging private consultants to assist it in collecting and analyzing communications intelligence (Bamford 2008; Priest and Arkin 2011, pp. 189–90). Whistleblower Edward Snowden was one of them, having been previously employed by a major government contractor, Booz Allen. In most cases of state or state-sponsored cybercrime, the degree of explicit state involvement remains obscure; this should come as no surprise, as plausible deniability is a valued condition for all criminals, state or private, terrestrial or cyber.

Most types of cybercrime can be committed by individuals or by organizations, although the latter can be capable of activities on a grander scale. At the extreme, operations as complex and sophisticated as Olympic Games, the cyber espionage activities of the People's Liberation Army, and the Prism project of the US National Security Agency require very many hands.

State cybercrime activity tends to involve espionage, surveillance, and destruction or degradation of an adversary's systems. It tends to be motivated by the need to identify and to neutralize perceived national security threats, or to give one's nation a strategic advantage. Conventional organized cybercrime activity is more diverse, and depending on crime type, tends to involve a wider variety of objectives, from financial gain, to sexual gratification, to political protest, to notoriety. Both state and private cybercrime may produce adverse unintended consequences for the perpetrator; state and state-sponsored cybercrime may, ironically, cause harm to a state's own national security.

Iatrogenic security threats

Organizations sometimes inflict damage upon themselves, most commonly through strategic misjudgment. Commonly, they fail to adapt to a changing environment. In the commercial world, businesses may fail to be attuned to market signals. Paoli (2003) describes the lack of adaptability on the part of some traditional mafia groups: Among the factors she identifies are the inward focus on family, village, and neighborhood; the inability or unwillingness to broaden the pool of new recruits; and the consequent failure to exploit emerging market opportunities such as the growing illicit arms trade.

Organizations may also suffer from overextending themselves. Business history is littered with the remains of companies whose directors, blinded by hubris or emboldened by previous successes, sought to expand too rapidly. The overreach of organized crime may also backfire. The assassinations of Sicilian investigating judges Giovanni Falcone and Paolo Borsellino in 1992 were more than just mafia "hits." The massive explosions themselves constituted a political statement and were obviously intended to discourage further investigative activity. Ironically, these killings had the effect of provoking widespread revulsion to organized crime and of fostering public support for increased law enforcement powers (Cowell 1992; Paoli 2003, p. 204).

The decisions of governments to go to war may be similarly flawed. Far from enhancing their own national security, Napoleon's invasion of Russia and Japan's attack on Pearl Harbor each led to disastrous consequences for the aggressors. The second Iraq war cost the United States one trillion dollars, over 4,000 military personnel killed in action, and considerable reputational capital. It incited further Al Qaeda activism and also left Iran unrestrained by what had been a formidable adversary.

Some of the more significant cyber threats to national security are also self-inflicted. States that commit cybercrime may themselves be weakened as a result, especially when their activities come to public attention. States that present themselves as paragons of virtue, only to be found to have been engaged in criminal activities, may see their moral authority eroded. Hypocrisy tends to be inconsistent with leadership. The state that does not practice what it preaches may lose legitimacy, both domestically and internationally. This appears to have been the case recently with the United States. In 2013, the US government became increasingly strident in voicing its displeasure with the government of the People's Republic of China regarding their programs of economic espionage. President Obama was to raise the issue during his June 2013 summit meeting with President Xi Jianping. However, Obama was upstaged by whistleblower Edward Snowden, who disclosed the Prism program and thereby revealed that the United States was itself engaged in a massive program of telecommunications interception.

Once it becomes public knowledge, the modus operandi of state cybercrime may well enhance the capacity of ordinary cybercriminals and other states. The escaped "Stuxnet" virus, reportedly used in the operation against Iranian nuclear enrichment facilities, is now available for use or further refinement by rogue states and cybercriminals, organized or acting alone.

The potential for "conventional" organized cybercrime to threaten national security is real. Its likely impact, however, is by no means constant across all nations. All else equal, states with weak economies and those lacking in social cohesion face a greater security risk than robust, developed nations. Relatively disadvantaged nations may be less able to afford the IT security infrastructure to protect state and private information systems (such as they are), particularly those supporting financial services. They may also lack the regulatory and enforcement resources to prevent and control cybercrime. But the more fortunate nations themselves may be attractive targets to cybercriminals. They contain a larger number of targets, both symbolic and material.

In addition, state or state-sponsored cybercrime may inspire imitative criminal behavior by other states and individuals alike. Leadership can function in a wholly unsuitable manner. Moreover, states can also provide imitators with a vocabulary of extenuation, a means of rationalizing or neutralizing criminal acts. The appeal to a higher purpose, the argument that the end justifies the means, will be particularly resonant with other states. History has shown that many states have pursued objectives which they at the time regarded as honorable but which turned out to be terribly perverse.

The full consequences of Operation Olympic Games are uncertain, as neither the victims nor the perpetrators are inclined to discuss it. With knowledge of the operation now widespread, and some of the software elements in the public domain, there are individuals and organizations who may follow the example of perpetrators and engage in cyber attacks for their own purposes. The ultimate consequences of this potential turn of events are unpredictable. Recent electronic attacks against US financial institutions and Saudi oil facilities may represent two examples (Perlroth 2012; Perlroth and Hardy 2013).

Conclusions

Not all organizations acting illegally in cyberspace may be regarded as threats to national or international security. Some of their activity is unquestionably annoying, offensive in the extreme, and/or harmful. There are online activities which, if writ large, might conceivably weaken the integrity and economy of states and thus come to be regarded as security threats. Online paedophilia, no matter how heinous or distasteful one might regard it, does not occur on a scale at which a nation's economy or social fabric is damaged. Nor is it likely ever to do so. The software and entertainment industries in the United States have sought to make the case that information piracy costs the domestic economy millions of dollars in lost profits and has a chilling effect on entrepreneurialism and creativity worldwide. However, the US economy is sufficiently diverse and vigorous that its future strength will depend on other factors. Indeed, one could argue that piracy allows citizens of less developed countries to benefit from access to products that they would otherwise be unable to afford. By contrast, major, persistent industrial or political espionage on a wider scale may well threaten the security of the target state. Industrial espionage can weaken the international competitiveness of a company or of an industry sector. When the information in question relates to weapons systems, the security implications are obvious. Theft of credit card and banking details, were it to occur on a larger scale, would certainly impede commercial activity, with corresponding harm to a state's economy. For the time being, however, online fraud appears manageable. Technologies of information security continue to be refined, and tech-savvy members of the public, at least in developing countries, are able to protect themselves. Online banking and e-commerce continue to thrive.

In less developed countries, however, where users are just entering the digital age, lack of awareness of cybersecurity may enhance the vulnerability of individuals and of organizations in both the public and private sectors. To the extent that the threat of cybercrime impedes the development of electronic commerce, it could contribute to the weakening of a nation's economy, and by extension, its security.

As far as organizations are concerned, it appears that the greatest threats to national security are posed by states themselves, either singlehandedly or in collaboration with skilled non-state actors. Resources available to the strongest states are indeed formidable, and we have seen them used with considerable effect. The activity reported to have led to the destruction of Iranian centrifuges would certainly be defined by authorities of that country (or those of any other nation on the receiving end of a similar attack) as a security threat. So too would denial of service attacks that degrade government information systems or those servers that support critical infrastructure. Indeed, nations on the receiving end of such attacks might be inclined to regard them as acts of cyber war (Turns 2012).

The activities of Anonymous, for example, eavesdropping on a conference call between law enforcement agents and attempts to shut down the website of the Central Intelligence Agency, have certainly been found annoying by authorities in the United States. On the scale at which they were conducted, activities of this kind were certainly embarrassing but would not constitute harm to national security. However, if undertaken persistently or on a larger scale, they might convey the impression of state nonchalance at best, and incompetence at worst, and thereby invite imitation from many quarters. The *potential* for significant harm is therefore real, if perhaps somewhat distant. The threshold of objective threat would appear to be a function of the scope and intensity of criminal activity, on the one hand, and the resilience of the state and its infrastructure on the other.

Other security threats may be self-inflicted by states as the result of their own cybercriminal activity. The threat intensifies when the illegality in question reaches public attention. This is not to suggest that the only crime is "getting caught." Any illicit activity on the part of the state runs the risk of detection and exposure, whether by insiders or external adversaries.

If organized cybercrime were to occur on a greater scale, it could lead to a weakening of trust in major public or private institutions. The fundamental question for our purposes is at what point does the nature and scale of online criminal activity (organized or otherwise) begin to constitute a security threat. National security is not a binary variable. The question is not "are you, or are you not, secure?" but rather "might a specific circumstance contribute to an enhancement or diminution of security?"

Bibliography

Bamford, James (2004) *A Pretext for War,* New York: Doubleday.

Bamford, James (2008) *The Shadow Factory: The NSA from 9/11 to the Eavesdropping on America,* New York: Random House.

Bradsher, Keith (2006) "North Korean Ploy Masks Ships Under Other Flags," *New York Times,* 20 October, http://www.nytimes.com/2006/10/20/world/asia/20shipping.html?pagewanted=print (accessed 25 July 2014).

Choo, Kim-Kwang Raymond and Grabosky, Peter (2014) "Cyber Crime," in *The Oxford Handbook of Organized Crime,* L. Paoli (ed.), New York: Oxford University Press.

Coleman, Gabriella (2011) "Anonymous: From the Lulz to Collective Action," *The New Everyday,* 6 April, http://mediacommons.futureofthebook.org/tne/pieces/anonymous-lulz-collective-action (accessed 25 July 2014).

Cowell, A. (1992) "Sicilians Jeer Italian Leaders at a Funeral Protest," *The New York Times,* 22 July, http://www.nytimes.com/1992/07/22/world/sicilians-jeer-italian-leaders-at-a-funeral-protest.html (accessed 25 July 2014).

Cressey, Donald R. (1972) *Criminal Organization: Its Elementary Forms,* London: Heinemann Educational Books.

Denning, Dorothy (2001) "Activism, Hacktivism, and Cyberterrorism: The Internet as a Tool for Influencing Foreign Policy," in *Networks and Netwars: The Future of Terror, Crime, and Militancy,* J. Arquilla and D. F. Ronfeldt (eds.), Santa Monica, CA: Rand Corporation.

Donohue, Laura (2013) "NSA Surveillance May be Legal – But It's Unconstitutional," *The Washington Post*, 21 June, http://articles.washingtonpost.com/2013-06-21/opinions/40110321_1_electronic-surveillance-fisa-nsa-surveillance (accessed 25 July 2014).

Federal Bureau of Investigation (FBI) (2012) "Press Release: Another Cyber Fraud Defendant Charged in Operation Aching Mules Sentenced in Manhattan Federal Court," 23 March, http://www.fbi.gov/newyork/press-releases/2012/another-cyber-fraud-defendant-charged-in-operation-aching-mules-sentenced-in-manhattan-federal-court (accessed 25 July 2014).

Gidda, Miren (2013) "Edward Snowden and the NSA Files – Timeline," *The Guardian*, 26 July, http://www.theguardian.com/world/2013/jun/23/edward-snowden-nsa-files-timeline?INTCMP=SRCH (accessed 25 July 2014).

Glenny, Misha (2011) *Dark Market*, New York: Knopf.

Grabosky, Peter (1995) "Using Non-governmental Resources to Foster Regulatory Compliance," *Governance: An International Journal of Policy and Administration* 8(4): 527–50.

Graham, William R, Jr (2000) "Uncovering and Eliminating Child Pornography Rings on the Internet: Issues Regarding and Avenues Facilitating Law Enforcement's Access to 'W0nderland'," *Law Review of Michigan State University – Detroit College of Law*, 457–84.

Greenwald, Glenn (2014) *No Place to Hide: Edward Snowden, the NSA, and the US Surveillance State*, New York: Metropolitan Books.

Greitens, Sheena (2012) "A North Korean Corleone," *New York Times*, 12 March, http://www.nytimes.com/2012/03/04/opinion/sunday/a-north-korean-corleone.html?pagewanted=all (accessed 25 July 2014).

Halstead, Boronia (1998) "The Use of Models in the Analysis of Organized Crime and Development of Policy," *Transnational Organized Crime* 4(1): 1–24.

Information Warfare Monitor (2010) "Shadows in the Cloud: Investigating Cyber Espionage 2.0," http://www.nartv.org/mirror/shadows-in-the-cloud.pdf (accessed 25 July 2014).

Lee, Jennifer (2002) "Pirates on the Web, Spoils on the Street," *New York Times*, 11 July, http://www.nytimes.com/2002/07/11/technology/pirates-on-the-web-spoils-on-the-street.html?pagewanted=all&src=pm (accessed 25 July 2014).

McIllwain, Jeffrey (2005) "Intellectual Property Theft and Organized Crime: The Case of Film Piracy," *Trends in Organized Crime* 8(4): 15–39.

Mandiant Intelligence Center (2013) *APT1: Exposing One of China's Cyber Espionage Units*, http://intelreport.mandiant.com/ (accessed 25 July 2014).

Markoff, John (2009) "Vast Spy System Loots Computers in 103 Countries," *New York Times*, 28 March, http://www.theglobeandmail.com/technology/meet-the-canadians-who-busted-ghostnet/article1214210/?page=all (accessed 25 July 2014).

Markoff, John and Barboza, David (2010) "Researchers Trace Data Theft to Intruders in China," *New York Times*, 5 April, http://www.nytimes.com/2010/04/06/science/06cyber.html?pagewanted=all (accessed 25 July 2014).

Martin, Brian G. (1996) *The Shanghai Green Gang: Politics and Organized Crime 1919-1937*, Berkeley, CA: University of California Press.

Morselli, Carlo (2009) *Inside Criminal Networks*, New York: Springer.

New York Times (2013) "More Fog from the Spy Agencies," *New York Times*, 31 July, http://www.nytimes.com/2013/08/01/opinion/more-fog-from-the-spy-agencies.html?emc=edit_tnt_20130731&tntemail0=y (accessed 25 July 2014).

Olson, Parmy (2012) *We Are Anonymous: Inside the Hacker World of LulzSec, Anonymous, and the Global Cyber Insurgency,* New York: Little Brown.

Paoli, L. (2003) *Mafia Brotherhoods: Organized Crime Italian Style,* New York: Oxford University Press.

Perl, Raphael (2007) "Drug Trafficking and North Korea: Issues for US Policy," Washington, DC: Congressional Research Service, http://fas.org/sgp/crs/row/RL32167.pdf (accessed 20 May 2013).

Perlroth, N. (2012) "In Cyberattack on Saudi Firm, US Sees Iran Firing Back," *New York Times,* 23 October, http://www.nytimes.com/2012/10/24/business/global/cyberattack-on-saudi-oil-firm-disquiets-us.html?pagewanted=all (accessed 25 July 2014).

Perlroth, N. and Hardy, Q. (2013) "Bank Hacking Was the Work of Iranians, Officials Say," *New York Times,* 8 January, http://www.nytimes.com/2013/01/09/technology/online-banking-attacks-were-work-of-iran-us-officials-say.html (accessed 25 July 2014).

Priest, Dana and Arkin, William (2011) *Top Secret America: The Rise of the New American Security State,* New York: Little Brown.

Russell, Gabrielle (2008) "Pedophiles in Wonderland: Censoring the Sinful in Cyberspace," *Journal of Criminal Law and Criminology* 98(4): 1467–500.

Schmidt, M. S. and Sanger, D. E. (2014) "5 in China Army Face US Charges of Cyberattacks," *New York Times,* 19 May, http://www.nytimes.com/2014/05/20/us/us-to-charge-chinese-workers-with-cyberspying.html# (accessed 25 July 2014).

Sanger, David (2012) *Confront and Conceal: Obama's Secret Wars and Surprising Use of American Power,* New York: Crown Publishers.

Sanger, D., Barboza, D. and Perlroth, N (2013) "Chinese Army Unit Is Seen as Tied to Hacking Against US," *New York Times,* 18 February, http://www.nytimes.com/2013/02/19/technology/chinas-army-is-seen-as-tied-to-hacking-against-us.html?pagewanted=all (accessed 25 July 2014).

Turns, David (2012) "Cyber Warfare and the Notion of Direct Participation in Hostilities," *Journal of Conflict Security Law* 17(2): 279–97.

Standing, Andre (2003) *The Social Contradictions of Organized Crime on the Cape Flats,* Institute for Security Studies Occasional paper 74, Pretoria, South Africa: Institute for Security Studies.

US Department of Justice (2002) "Warez Leader Sentenced to 46 Months," 17 May, http://www.justice.gov/criminal/cybercrime/press-releases/2002/sankusSent.htm (accessed 25 July 2014).

Urbas, Gregor (2006) "Cross-national Investigation and Prosecution of Intellectual Property Crimes: The Example of 'Operation Buccaneer'," *Crime, Law and Social Change* 46: 207–21.

Wakeman, Frederick Jr (1995) *Policing Shanghai, 1927–1937,* Berkeley, CA: University of California Press.

Weiner, Tim (2007) *Legacy of Ashes: The History of the CIA,* New York: Doubleday.

Williams, Phil (2001) "Transnational Criminal Networks," in *Networks and Netwars,* J. Arquilla and D. Ronfeldt (eds.), Santa Monica, CA: Rand Corporation.

Zetter, Kim (2014) *Countdown to Zero Day: Stuxnet and the Launch of the World's First Digital Weapon,* New York: Crown.

6

Internet Fraud in Hong Kong: An Analysis of a Sample of Court Cases

Laurie Yiu-Chung Lau

Introduction

This chapter analyzes a selection of court cases concerning Internet crimes, particularly those involving Internet banking fraud. The rule of law is a core value of the government in Hong Kong, which seeks to provide a secure environment in which both individuals and organizations can transact business both online and offline. This chapter provides a snapshot of how Hong Kong's Internet governance regime operates by examining a sample of Internet fraud cases from the mid-1990s. It shows how the rule of law has helped Hong Kong to achieve its place as an economic hub in Asia.

Internet banking fraud has specific features in Hong Kong that differ from other centers of economic crime. These were explained in an interview conducted with a specialist technology crime prosecutor from the Hong Kong Department of Justice in early 2006:

> Hong Kong is different from many countries in the West. For example, the USA is a large country, therefore, it is very popular for people in the States to buy things over the Internet whereas Hong Kong is geographically a small place – just spend two minutes or five minutes and you are in a shop.... In the States in a lot of cases, they do it for fun. For example, hacking into the FBI [website] just to deface it for one minute. If that particular guy managed to deface the FBI website, then he can show off to his friends.... I haven't seen that in Hong Kong. For e-fraud ... they attempt to sell one item to a hundred people, get the money and then they're gone. [In Hong Kong] they are really [after] serious money when they commit that offence. I think that's the main difference.

Although a great deal of money, time, and skill is spent on investigating Internet-enabled economic crime in Hong Kong, not all cases result in a trial, even though most cases are likely to have been thoroughly investigated. Even fewer cases reach the Appeal Court. Nonetheless, appellate decisions provide one of the few sources of information on cases that have been investigated and how courts dispose of them. The present study of Internet banking fraud cases that have gone to trial and then been subject to appeal reveals that the policing of economic Internet crimes in Hong Kong is influenced by a number of factors. These include Hong Kong's political structure as a non-democratic, semi-authoritarian city state; the presence of a free market economy; the legal environment; local technological and financial astuteness and the increase in "mainlandization" after 1997. The present research also provides evidence of the extent of liaison between the banks and the police and the capability of the police to investigate complex cases of Internet banking fraud, especially international cases. The use of the *Computer Crime Ordinance 1993* (CC Ord 1993) was also examined, and consideration was given to whether the rule of the law was applied equally or selectively to a few high-profile cases. The impact of mainlandization of Internet banking crime was also examined in addition to how the courts approach sentencing in these kinds of cases.

Methodology and sample

The present study involved an analysis of a sample of appellate decisions involving Internet banking fraud that came before the Hong Kong Court of Appeal between 1999 and 2012. Relatively few cases of Internet banking fraud in Hong Kong come before the courts, and those that do are likely to be the tip of the iceberg. According to a chief inspector of the Technology Crime Division of the Hong Kong Police, interviewed in early 2006, 'a lot of Internet-related crimes happen, such as online shopping fraud or auction fraud and e-banking theft... but the victims in such cases often do not want to report the crimes to the authorities' (Personal Communication, 18 March 2006). As a result, these cases are part of the "dark figure" of under-reported Internet-related crime. Sometimes such cases are dealt with by private investigators who report to the police only as a last resort. According to a regional director for security of a global credit card company, when interviewed in early 2006:

... most of the private sector investigators, if they come across any computer crime cases ... really want to refer the case to the police but that will take some time, so they prefer to ... conclude the cases by themselves ... until they think that ... they need to ask law enforcement people to prosecute. Otherwise, I would say most of the cases are not reported (Personal Communication, 18 March 2006).

Accordingly, the present study is based on only a small sample of the most serious cases that have been referred for prosecution and then been subject to appeal. Case selection was undertaken using the methodology of Wing Lo and Ngan (2009). This involved conducting a database search of appellate decisions using the following search terms: Internet banking fraud, Hong Kong, Computer Crime Ordinance 1993, mainlandization, political structure, semi-authoritarian city state, free market economy, legal environment, local technological and financial savviness.

The sample 15 cases were selected from the Westlaw Hong Kong case law database of appellate decisions between 1 January 1999 and 30 June 2012 (13.5 years), well after the Computer Crime Ordinance came into force in 1993.

In addition to the case analysis, the present study also involved consultations with key figures involved in the investigation and prosecution of Internet banking fraud in Hong Kong. Five interviews were conducted with police and other relevant individuals between January and June 2006. Undertakings as to confidentiality were given and, accordingly, names of those interviewed and their organizations have been suppressed.

General case characteristics

All 15 cases were on appeal from the Hong Kong Magistrates' Court. In almost all cases, the motive was monetary gain (except Case 1 where the motive was political). Almost all cases (Cases 1–5 and Cases 7–15) were initially investigated by the police under the Computer Crimes Ordinance 1993 but then prosecuted as a different offense. Apart from Case 6, all went to trial. In Case 6, the appellant appeared before a police internal disciplinary tribunal before appealing via judicial review. All the defendants used the Internet to commit the offenses, and none of the defendants were related to their victims. The main features of the 15 cases are presented in Table 6.1.

Table 6.1 Principal characteristics of cases in the sample

Case No.	Initial investigation under CC Ord 1993	Actual offense indicted	Advantage Monetary	Non-Monetary	Connection with mainland China	Internet enabled offense	Victim unrelated to accused*	Appeal present	Other offenses charged
1	yes	CC Ord 1993	no	yes	no	yes	yes	yes	
2	yes	CC Ord 1993	yes	no	no	yes	yes	yes	copyright
3	yes	CC Ord 1993	yes	no	no	yes	yes	yes	theft
4	yes	CC Ord 1993	yes	no	no	yes	yes	yes	
5	yes	Conspiracy to defraud	yes	no	yes	yes	yes	yes	
6	Police Internal Tribunal	Conduct calculated to bring public service into disrepute	yes	no	no	yes	yes	Judicial Review	
7	yes	Dealing in stolen property	yes	no	no	yes	yes	yes	Money laundering
8	yes	Handling stolen goods	yes	no	no	yes	yes	yes	
9	yes	Dealing in stolen property	yes	no	no	yes	yes	yes	
10	yes	Theft	yes	no	yes	yes	yes	yes	
11	yes	Insider dealing	yes	no	yes	yes	yes	yes	
12	yes	Theft	yes	no	yes	yes	yes	yes	
13	yes	Dealing in stolen property	yes	no	yes	yes	yes	no	
14	yes	Fraud, theft, and soliciting advantage	yes	no	yes	yes	yes	no	Corruption
15	yes	Dealing in stolen property	yes	no	yes	yes	yes	no, not guilty at the trial	

*Absence of blood relationship or close relative

Case charges and outcomes

All the cases were initially investigated for offenses under the Computer Crime Ordinance 1993. In Case 9, the defendant pleaded guilty in the District Court for an offense contrary to section 25 of the Organised and Serious Crimes Ordinance, Cap. 455. In Cases 5–15, the defendants were initially investigated by the authorities for offenses under the Computer Crime Ordinance 1993, such as unauthorized access to the Internet. However, at trial, the actual indictments were for other offenses, some of which were far more serious. For example, in Cases 7, 9, 13, and 15, the indictments at court, which included dealing in stolen property, differed from the initial police charges.

In Cases 7, 9, and 13, the accused were found guilty. In Case 15, the court concluded, "not without reluctance," that there was reasonable doubt, and the accused was acquitted of dealing in property known or reasonably believed to represent the proceeds of an indictable offense which the prosecution failed to prove beyond reasonable doubt.

All cases involved the use of the Internet and gaining unauthorized computer access, and all but one involved personal monetary gain (Cases 2–15). In Cases 2, 3, 4, and 9, the defendants pleaded guilty at the trial but later appealed their sentences. In Cases 13 and 14, the defendants were found guilty after a trial at the District Court. In Case 15, the defendant was found not guilty at trial. Details of the case outcomes by gender of the accused are shown in Table 6.2 from which it appears that only three out of the 26 were female (12%).

In terms of case outcomes, almost all (13 out of 15) of the cases involved appeals, mainly against sentence, except for Cases 13 and 14. Almost all of the accused (24 out of 26 defendants) either pleaded guilty or were found guilty at trial, except for Case 15, in which the accused was found not guilty at trial. There was a high rate of custodial sentences being imposed, with 18 out of the 24 defendants found guilty receiving a custodial sentence (75%). The exceptions were Case 1 (community service), Case 6 (judicial review upheld his dismissal), Case 8 (fine of HK$5000); and Case 14, where the conviction was overturned on appeal.

Case complexity

Cases 5, 11, 12, 13, 14, and 15 were all complex cases. For example, Case 11 included several complex crimes that involved the Hong Kong Monetary Authority (HKMA) and the Securities and Futures Commission (SFC). The accused were prosecuted by the SFC for insider dealing using

Table 6.2 Case outcomes by gender

Case number	Number of defendants	*M/F	Guilty plea / factual decision	Custodial sentence (number of defendants)
1	1	M	1	None
2	3	3 M	3	3
3	1	M	1	1
4	1	M	1	1
5	5	5 M	5	5
6	1	M	Disciplinary tribunal found against the defendant	Not applicable
7	1	M	1	1
8	1	M	1	Reduced to fine on appeal
9	1	M	1	1
10	2	2 F	2	2
11	5	2 F 3 M	5	2
12	1	M	1	1
13	1	M	1	1
14	1	M	1	Custodial sentence imposed on appeal
15	1	M	0**	none
Total	26		24 (92.3%)	18 (75.0%)

*Male=M, Female=F. ** One accused was found not guilty following judicial review.

an Internet bank account to deal in shares. The five defendants in the case were all closely related. The first defendant was employed as a senior executive (vice president) in an investment bank. He was handling a project relating to the privatization of a company listed on the Hong Kong Stock Exchange and possessed important price-sensitive information. He deliberately leaked the sensitive information to his brother, the third defendant, who then passed it on to the other relatives. Based on the information gained from the third defendant, the second, fourth, and fifth defendants bought large numbers of shares via their HSBC Internet bank accounts (over 1.6 million shares at between HK$1.58 and HK$1.60). They did this within a 15-minute period. Two weeks later, the privatization of the listed company was announced, with a cash offer of HK$1.80 per share. The defendants thus profited from their share holdings. The authorities were alerted to the case because the volume of shares that were traded within the 15 minutes constituted 93 percent of all the shares traded in the company on the stock exchange that day (6 June 2006).

This case highlights the fact that the Hong Kong authorities keep a watchful eye on the Hong Kong financial markets and the banking system for abnormalities such as this. It also highlights the fact that Internet bank accounts in Hong Kong are not only used for simple personal money transactions but can also be used for the online trading of stock and shares involving large sums of money.

By prosecuting this case, the authorities sent a clear message to others contemplating committing this type of crime. These actions enable the authorities to help protect the integrity of the Hong Kong Internet banking system and the stock market. As the judge stated in *HKSAR v Ma Hon Kit Sammy & ORS* (2010, at pp. 4–5), "...the electronic banking system and a 'clean' stock market are vital for Hong Kong." This landmark case was widely reported in the media. Martin Wheatley (n.d.) of Security and Futures Commission noted that this was the first insider dealing case to involve a prison sentence since insider dealing was made a criminal offense under the Securities and Futures Ordinance in 2003. Ma was given a sentence of 26 months and his girlfriend, and co-accused, 12 months, while his nephew received 200 hours of community service and a fine of HK$17,000.

Protecting Hong Kong's status as one of the world's leading capital markets is an important priority for the government and the police. As Hong Kong is one of the world's leading capital markets, the stock exchange and the related financial industries provide a huge source of income for the government in terms of tax. In recent years, the Hong Kong Stock Exchange has also begun to serve as a hub for mainland Chinese companies seeking to raise capital and a channel for listing on the open market. Today, more than 50 percent of the stocks listed in Hong Kong are for mainland Chinese companies. In 2012 alone, as the report by Deloitte Touche Tohmatsu (2012) showed, 62 mainland companies were newly listed on the Hong Kong stock exchange, raising over HK$90 billion. Therefore, this case confirms that the courts take a serious view of activities that have the potential to damage Hong Kong's economy and reputation. The case also provides evidence of the role of the HKMA in detecting suspicious transactions.

Cases 5, 10, 12, and 13 were also complex cases involving large amounts of money and a link to mainland China. In Case 5, *Re Chin Kam Chiu and 4 others* (2005), all the defendants were found guilty in the Court of First Instance of a single count of conspiracy to defraud. At trial, the first defendant was sentenced to 6.5 years, the second defendant to 4.5 years, the third defendant 5.5 years, and the fourth and fifth defendants to 4.5 years. They all appealed their conviction and sentences.

Four of the appellants were officers of the Sin Hua Bank (a deputy general manager, a deputy manager in the credit department, an assistant general manager of the bills department, and a senior manager in a sub-branch). In 2001, the bank merged with and formed part of the Bank of China, whose headquarters are in Hong Kong. The defendants used their positions as bank officers to apply for 25 letters of credit (L/Cs), amounting to HK$1.8 billion, using false documents, such as false cargo receipts and false bills of lading. One of the appellants was the bank's in-house accountant. He signed the applications and used the Sin Hua Bank's Internet connected banking system to apply to the bank's Tsuen Wan bills center, which issues credit documents, for the L/Cs.

This case was also connected to mainland China. As the prosecution stated, "...these 25 letters of credit, although they were opened in Hong Kong through the Tsuen Wan Bills Centre, and although the applicants' accounts were dealt with at the Castle Peak Road sub-branch of the Sin Hua Bank, these particular L/Cs were channelled through Shenzhen and drawn on the Shenzhen Branch credit lines of the Sin Hua Bank." Although one defendant suggested that the Shenzhen branch was a separate legal entity, the court said there was no doubt that the defense were aware of the prosecution's allegations that the Shenzhen branch was an integral part of the Sin Hue Bank group. According to the court documents, the bank was a limited liability company registered in mainland China with its headquarters in Beijing. All the issued shares in the bank were owned by the Chinese government. The court added that the fact the Shenzhen branch was outside the HKMA's control was unrelated to the real issue in the case, as there was unity of management and control between the Hong Kong and Shenzhen branches.

The case came to light in April 1998, when the HKMA expressed concern over the bank's exposure to the Keen Lloyd Group (which was owned by the appellants and operated in mainland China). The exposure of HK$1.8 billion was approximately 25 percent of the bank's capital. At the same time, the HKMA also expressed concern that the bank's credit line to the Keen Lloyd Group had increased 50 percent since 1997. The HKMA requested that the bank reduce its exposure to the group. Eventually, the case was referred to the police and the ICAC for investigation. The defendants' appeals against their convictions and sentences were dismissed.

This case also illustrates something mentioned by Linklaters (2010) in their overview of the role of the FSC in investigating and prosecuting insider dealing. They said that over the past two years, they had

… seen a paradigm shift in the approach adopted by the Securities and Futures Commission (SFC) in its investigations, with use being made of previously untested statutory powers to freeze assets, compel production of documents and record interviews with employees. Insider dealing has been top of the SFC's agenda throughout. However, it was not until 2008 that the SFC brought its first criminal prosecution for insider dealing (some five years after the offence was added to the statute books). The convictions of five individuals in the case of HKSAR v Sammy Ma Hon Kit and others (Criminal Appeal no. 148 of 2009), one of whom was a banker who passed inside information to his girlfriend and family in relation to a deal on which he was advising, were secured over a year later. Fines were imposed on each and sentences of imprisonment were passed in respect of the two main parties. Last month, Hong Kong's Court of Appeal refused leave to appeal against the convictions of two of the five. The Court of Appeal was satisfied that the irresistible inference to be drawn from the trading activities of these family members was that they had received inside information from the banker (Linklaters 2010, p. 1).

Linklaters has also pointed out in the past that many such cases had previously been dealt with via the civil – not criminal – legal process.

Trans-jurisdictional proceedings

Cases 5, 10, 12, 13, 14, and 15 all involved a link between Hong Kong and mainland China, while Case 12 involved Hong Kong, mainland China, and a minor connection with the United States (US). Case 14 stretched between the US, mainland China, Hong Kong, and Europe; while Case 15 involved Hong Kong and Dubai. Nonetheless, as the head of the Technology Crime Division (TCD) confirmed when interviewed in early 2006, even though these cases were trans-jurisdictional, the perpetrators were still within the reach of Hong Kong law:

> In Hong Kong we are trying to do something on the crime jurisdiction ordinance (CJO), so that if a criminal in HK is doing something wrong overseas HK law and jurisdiction can try the crime. On the other hand, if somebody is doing something wrong in HK from overseas the law can also try this person too. However, we still have quite a long way to go, if you look at the European Cybercrime Convention, the Interpol best practice, there were a lot of big principles, but

because those principles are too big, nobody really follows it (Personal Communication, 11 April 2006).

He went on to say:

> The way we look at the moment, a lot of those Internet commercial crimes that occur here are not posted in HK; it was somewhere else. If you look at HK geographically, because it is so small, if anything comes up, we can close it up within one day. Last Friday, there was a report of an advance fees scam in Kowloon, on Saturday morning I sent my men down there to seize the server and closed it down. We can do that. However, a lot of the international fraudsters ... are not locating their servers in Hong Kong, they locate their servers somewhere they think is safe and the police there [are] not up to it. So, they host their web around the world. So, a lot of these activities are international, but it can also affect HK...you might not know that one case loss HK$1.3m on e-banking, however, we stop it in time, so the loss zero... (Personal Communication, 11 April 2006).

The head of the Technology Crime Division interviewed in early 2006 also explained some steps that might be needed before solving such a case:

> There are a lot of important issues, when you are looking at this kind of thing. First of all, [is] speed: how fast we can obtain the electronic data, such as the computer record, IP record? It is very difficult for us to obtain record when it is trans-border. Often it is the time – we get all information but it is too late. Put it this way, we settle down several issues, first of all, technical issues in getting the evidence, secondly, after getting the record from whoever that is (the person) and assistance from whoever that is. Can we get the computer forensic evidence to prosecute or prove against the crime – that is not that big a problem in Hong Kong, because we can take our time. The longest computer forensic examination took about nine months, involving a case of the computer network, ISP, 80 computers and two Internet servers. We just closed them down and seized it, brought them all back here and rebuilt the system in here at the lab. Then, we were examining it for nine months. After the first three months, we said we can't handle it; we asked Microsoft to assist us. So after nine months, we got everything we wanted, so we prosecuted the persons involved... if the case occurred locally, then we can easily do it, but if it is something on overseas, then we will have problem (Personal Communication, 11 April 2006).

The mainland China–Hong Kong connection

Seven out of the 15 cases in the sample had a Hong Kong/mainland China link, namely, Cases 5, 10, 11, 12, 13, 14, and 15. This trend was confirmed by an Internet banker interviewed in early 2006 who said:

> ... Hong Kong is growing into... a city of China. We are still running a different system, unlike the mainland China, but economic activities are more likely to be related to the mainland Chinese. Now this kind of activity is happening within the region, what we call the greater China region, so the situation is getting more complicated... Internet or the electronic crime will grow and tackling it will be more difficult. Talking about, like, across region or going to mainland China, their system or systems for tackling this kind of thing are more problematic as well (Personal Communication, 3 March 2006).

For example, Case 14 involved a complex set of Internet transactions between mainland China and Hong Kong. The defendant was charged with fraud under the Theft Ordinance and with soliciting an advantage under the Prevention of Bribery Ordinance. He was employed as the technical director of a company that provided a platform for online payments and technical support for merchants in receiving payments from customers on the Internet. The company also cleared online credit card transactions relating to banks and financial institutions. These kinds of activities are encouraged by government initiatives on closer economic integration with mainland China.

The defendant was employed to set up a technical center in Zhuhai City, in Southern China, to handle programme development on the mainland, as it was said to be cheaper to set up an office in China. The office was set up in 2007, with three or four staff, in the name of Richard Chan, a Chinese citizen. The defendant admitted inflating the expenses of the Zhuhai center and pocketing the difference (563,318.23 renminbi) by depositing it into his bank account.

In this case, the defendant was alleged to have solicited an advantage and corruption involving commission fees from a trade partner in the US through email. The US company was an automatic clearing house that shared 25 percent of the profit generated by facilitating Internet payments. However, the defendant sent emails to the US company soliciting commission of 40 to 50 percent.

The District Court judge dismissed the corruption charge but convicted the defendant of fraudulently inflating an expenses claim. The case was investigated and brought to court by the ICAC. However, questions were raised about the admissibility of the evidence, as the ICAC officers were said to have conducted off-the-record interviews with the defendant. This became an important issue at appeal.

This highly complex case stretched between the US, Europe, Hong Kong, and Zhuhai City in China. Electronic crimes are often intertwined with other types of crime and often involve trans-border transactions that cross several jurisdictions. As such, they pose problems for the prosecuting authorities both in terms of their complexity and the cost of the cross-jurisdictional investigation. Moreover, it is often difficult to get a conviction in highly complex trans-border corruption cases that involve several countries, which often involve Internet technology. One difficulty for the prosecution is to prove the chain of evidence, as the evidence is likely to be enmeshed in traditional paper and electronic formats, which may be embedded in different electronic files on the Internet or in folders lodged in different jurisdictions. Moreover, the defendants can intentionally destroy electronic evidence by erasing files.

Prosecutions under Computer Crimes Ordinance 1993 together with other crimes

Cases 2, 3, 7, and 14 were prosecuted pursuant to the Computer Crimes Ordinance 1993 together with other serious crimes, such as corruption, theft, money laundering, and copyright. In all four cases the defendants were found guilty. These indicate how difficult it is to isolate Internet banking fraud as a separate offense, which may be one reason why the numbers for this specific offense seem so low. That is, the rates for Internet banking crime are "hidden" within other offenses.

Prosecutions originally under the Computer Crimes Ordinance 1993, but finalized as other offenses

In ten out of the 15 cases (Cases 5–15) (66.6%) the accused were initially charged under the Computer Crimes Ordinance 1993 but at trial they were charged with other offenses. Case 6, however, was not a criminal prosecution but a disciplinary hearing.

In Case 5, the accused was charged with conspiracy to defraud; in Case 7 the accused was charged with money laundering; in Cases 8, 9, 13, and 15, the accused were charged with handling stolen goods; in Cases 10 and 12, the accused was charged with theft; in Case 11 the accused was charged with insider dealing; in Case 14 the accused was charged with fraud, theft, soliciting an advantage, and corruption. These examples demonstrate that many of those who are initially charged with computer-enabled crimes were eventually charged with traditional crimes such as theft and handling stolen property. In these cases, although the defendants used the Internet to commit the crimes, they were not prosecuted specifically for Internet crimes.

Motivations

Only in Case 1 was the motive political rather than monetary. In Case 8, which involved stealing game points, it was not clear whether the accused simply wanted to play the casino game or wanted to use the bonus points to win cash.

The use of custodial sentences

The Hong Kong courts appear to take a strong line on all kinds of Internet and computer-related crime, not just those involving large amounts of money. In Case 2, for example, the third defendant (a teenager) was indicted for computer crimes (copyright offenses) which involved obtaining access to a computer with dishonest intent under the Computer Crime Ordinance 1993 and dishonest gain contrary to section 161 (1) (c) of the Crimes Ordinance. Two of the defendants who pleaded guilty in the Magistrates' Court later appealed their custodial sentences. Both of the defendants were teenagers who had strong family support, and their probation officers considered them suitable for a community service order. One had been studying at the Open University, and it was argued that a custodial sentence would disrupt his studying. However, dismissing the appeal, the appeal judge commented that a custodial sentence was neither wrong nor excessive, because "Whilst it is true that the number of prosecutions in respect of section 161 offenses is at present small, the damage which such offenses can cause should not be underestimated" (HKSAR v *Tam Hei Lun & ORS*-[2000] 3 HKC 745, Magistracy Appeal No 385 of 2000).

He added that "there have been less than ten prosecutions in relation to offenses under section 161 and in those circumstances it is most unlikely that the full range of crimes which would fall within section 161 would now be known or appreciated." The copying of copyrighted works and selling them to members of the public on the Internet commercially – even part time as a "cottage industry" – was a very serious offense.

The case in which the defendants were sentenced to imprisonment were indicative of the tough stance on crime control in Hong Kong and the belief that imprisonment is justified both as a punishment and a deterrent in cases of Internet-related "white collar" crime. For instance, in Case 3 the defendant received a custodial sentence after pleading guilty. He was charged with using a computer with dishonest intent to cause loss to another, contrary to section 161(1) (d) of the Crimes Ordinance. The defendant was convicted of theft in the Magistrates' Court for gaining unauthorized access to his employer's computer and defrauding the Internet service provider of HK$286.81. His appeal against his custodial sentence was dismissed despite a strong recommendation for community service. The appeal judge commented that although the offense only involved a small amount of money (HK$286.81), the defendant had betrayed the trust of the customer who had delivered the computer to him for servicing. The judge ruled that a custodial sentence would deter such conduct and that a period of disciplinary training "in a detention center" would be beneficial to the appellant.

This case demonstrates that the courts take a harsh approach to any illegal activity that threatens to damage Hong Kong's financial standing. In Cases 4 and 7 to 12, the defendants were also sentenced to custodial terms of various lengths. This high number of convictions confirms the view of a frontline police officer interviewed in early 2006, who said that

> a high number of [computer related crimes] cases which go to court end with conviction verdict. Even though they may appeal their verdict to a higher court, the higher court usually keeps the original guilty verdict (Personal Communication, 19 March 2006).

In contrast to the lower courts' strong custodial sentencing, in Case 1 the Court of Appeal overturned a prison sentence. Case 1 highlights a particular characteristic of Hong Kong's contemporary political situation. In this case, the defendant leaked details of the Secretary of Justice's medical record to two local newspapers and was prosecuted for unauthorized Internet access and fraud under the Computer Crime Ordinance 1993. He staged a "public interest" defense, stating that he had wanted to expose the fact the government had not told the truth about a key government official (the Secretary of Justice) being admitted to hospital for emergency surgery. The defendant's aim was to expose the newly formed Hong Kong Special Administrative Region government. By showing that the government had not been entirely honest about a key government official, the defendant's actions chimed with wider fears about the openness and transparency of the post-1997 regime. At the time, there was heightened public anxiety about the openness of the new HKSAR administration.

The defendant was found guilty under the Computer Crime Ordinance 1993 of unauthorized access to an Internet account and acting dishonestly and was sentenced to six months imprisonment. However, the appeal court judge in *HKSAR v Tsun Shui Lun* (1999) stated that "this is an exceptional case...I do think that in these peculiar circumstances, a term of six months' imprisonment is clearly manifestly excessive." The six-month sentence was reduced to 100 hours community service.

The circumstances surrounding this case were peculiar, as they suggested a political rather than a criminal motive. After being arrested, the defendant spent seven days in police custody before he appeared in the Magistrate's Court. However, the fact that the original sentence was overturned shows that the courts can act as a check on the power of the state, even where the case seriously embarrasses the government.

In most cases, the courts appear to be punishing the crime with little regard to the status of the offender. However, in Case 6 the status of the offender did come into play. This case involved a referral from the Transport Department to the police concerning suspicions that a third person was defrauding someone's Internet bank account. On a number of occasions, a check issued to the Transport Department had bounced just before the bank making the withdrawal tried to take the money from the defendant's bank account. In fact, the defendant (a police officer) was intentionally transferring the funds from his Internet bank account to another bank account just before the funds were withdrawn to cover the issued check, leaving insufficient funds to cover the check issued to the Transport Department.

In this case, however, instead of being prosecuted for a criminal offense, the defendant was dismissed following an internal police disciplinary tribunal. The tribunal found him guilty of misconduct. The officer sought a judicial review. However, the case was dismissed and the original decision of the disciplinary tribunal was upheld.

If no application for judicial review had been made, the case would never have reached the public domain. The matter would have been dealt with internally by the police, and the general public would never have become aware of the officer's misconduct. In following this course of action, the police themselves did not uphold the principle that everyone is equal before the law. Had it not been for the judicial review application, the matter would have bypassed the legal system altogether.

Therefore, it may be that other cases follow a similar route and add to the "dark figure" on unreported and/or under-reported crime in Hong Kong.

Because this case was originally dealt with behind the closed doors of a tribunal, it has the potential to feed into public fears about the erosion of the rule of law by public authorities in the wake of the 1997 transfer of sovereignty and Hong Kong's greater integration into the political structure of the mainland. Moreover, the public have had a longstanding suspicion about the independence and fairness of police internal disciplinary tribunals. As in the case of *Lam Siu Po* v *Commissioner of Police* (2009), the police disciplinary tribunal was found to be unfair and biased toward the police, leading to the police dismissal being quashed on appeal. However, the fact that the defendant believed that judicial review would provide him with justice suggests that at least some people still believe in due process under the rule of law.

Prosecutions involving minor Internet crimes

A further issue concerns the extent to which, and reasons why, relatively trivial and low-value offenses are prosecuted. In Cases 1, 3, and 8, the actual monetary value involved was either low or zero. Even so, the defendant in Case 1 received a six-month custodial sentence. In Case 3, the defendant was sentenced to a detention center, even though the monetary value of the crime was less than HK$290. In Case 8, the defendant who earned bonus points was fined HK$5,000. Although these are all relatively trivial cases, they would have taken valuable time and resources to investigate.

Mainlandization

Seven of the 15 cases involved transactions with mainland China. The majority of these cases involved large sums of money or were serious in other ways, such as corruption. In one case, Case 13, the corrupt mainland official identified in the case was beyond the reach of Hong Kong law. In cases such as this, corrupt mainland officials are dealt with not by the law but by internal party mechanisms. For example, Bloomberg Business Week (June 19, 2003) article reported that Liu Jinbao, the former head of the Bank of China in Hong Kong, faced charges in relation to irregular loans in China. He was recalled to Beijing and disciplined internally by the Communist Party for his shortcomings.

Equal application of the law

The cases in the sample suggest that "small fry" are prosecuted to the full extent of the law in Hong Kong as much as the "big fish." In other words, the police pursue all cases of banking fraud regardless of whether large or small amounts are involved, and the legislation is framed in such a way that it allows them to do so. The police also seem to be using the Computer Crimes Ordinance 1993 as a "catch-all," at least at the start of an investigation, before switching to other, often more serious charges later. It is easier for the police to build a stronger case after the initial charges have been laid. For example, in Case 1, a criminal law was used to prosecute a political act, which raises questions about the use of the law as a pretext for political policing.

The fact that the police will prosecute even minor offenses was confirmed by a chief inspector interviewed in early 2006 who said:

> Unlike other countries, other law enforcement agencies around the world, we do not set a minimum bar to take up the computer investigation, in terms of the monetary values etc... because investigation of the computer crime to HK police is still at a learning curve. So, even though the monetary value involved in a crime may be very low, we look at it as sort of learning process that our officer gains from it (Personal Communication, 19 March 2006).

However, an e-banker interviewed in early 2006 said his impression was that the police only seek to prosecute high-profile cases:

> ...so far they have sought high profile cases...like the one that was discussed at the AmCham seminar...but those are very individual case. I am not so sure whether the legal system or the prosecution in Hong Kong...like is sufficient enough to tackle this kind of case... but so far, there were some...like, more high profile cases exposed in Hong Kong...(Personal Communication, 26 March 2006).

Nonetheless, the minor offenses prosecuted in the present study may be taken as evidence that the rule of law is being equally applied. However, in addition to supporting the idea that everybody is equal before the law, these cases suggest that the police in Hong Kong take a tough approach to crime control and pursue everyone equally harshly. Accordingly, all offenders are likely to receive a custodial sentence even for minor crimes. In this sense, the Hong Kong criminal justice system leans toward crime control in favor of the prosecution rather than the defendant. This stance was modified with the establishment of the Bill of Rights in 1991, after the Tiananmen Square incident in 1989. Before that there was a tradition in the law of the presumption of guilt for a number of offenses. The Police and Criminal Evidence Act 1984, "PACE," is a code of practice that established the powers of the police to combat crimes while protecting the rights of the public. PACE set out the balance between the powers of the police and the rights and freedoms of the public. Between 1991 and 1997, the Bill of Rights made crime control secondary to due process. However, after the transfer of sovereignty in 1997, the Basic Law became the most important constitutional law document, for example, Basic Law instrument 9, which states, "The Basic Law of the Hong Kong Special Administrative Region is constitutional as it is enacted in accordance with the Constitution of the People's Republic of China and in the light of the specific conditions of Hong Kong. The systems, policies and laws to be instituted after the establishment of the Hong Kong Special Administrative Region shall be based on the Basic Law of the Hong Kong Special Administrative Region" overarching the national law of Hong Kong. Moreover, one of the requirements of the Basic Law is that Hong Kong must maintain its economic and political stability. Therefore, the "crime control" approach to policing apparent in the case samples may reflect an attempt to protect Hong Kong's reputation as a "clean" economy.

Technological and financial sophistication

A number of the defendants in the cases examined demonstrated a high level of technological and financial knowledge in conducting their crimes and exploiting new opportunities for gain. Case 8, for example, shows that this know-how, savviness, or sophistication is evident among ordinary youth as much as the commercial elite.

Police capability

A number of the officers interviewed in early 2006 stressed that the police were well trained and highly capable in investigating Internet banking fraud. Several also emphasized the fact that police liaised with ISPs, the Department of Justice, and banks:

> We [the police] have regularly scheduled meeting with the DoJ and ISPs to discuss various issues about computer related crimes in Hong Kong, including the current trend in computer related crimes, offender methods, both new and old methods. The discussions would

also involve how, and in what way, the ISPs can improve their side or the way of doing things, so that it can help the police to solve computer related crimes more quickly.

Case 7 reflects the capability of the Hong Kong police in investigating banking crime. In this case of Internet money laundering, bank transfers were made between Hong Kong, Taiwan, and South Korea. This international case also threatened Hong Kong's financial reputation. Nonetheless, the TCD was able to liaise with agencies in the other jurisdictions and together they successfully traced the computer evidence across several countries.

This kind of international liaison is part of the Hong Kong police's strategy for tackling Internet crime. As the head of the TCD interviewed in early 2006 stated, "we are up to our capability, in the sense of handling existing technology crime, i.e. hacking, website defacing, online deception and fraud." He also commented that, "since 2001, the Hong Kong police have been sent off to mainland China, to assist..." (Personal Communication, 11 April 2006).

The cases in this sample suggest that the police do have the capacity to conduct international investigations.

Bank/police liaison

The banks and the police both claim to liaise closely when it comes to Internet crime. As one respondent, a chief inspector in the TCD interviewed in early 2006, stated, "HK is a very small place, unlike the UK or US. It is quite easy for people to get together for meeting and exchange intelligence." However, an e-banker interviewed in early 2006 argued that

> the bank and the police like having very little communication, so on this kind of like computer crime situation, what we have done is, like, do it by ourselves, try to make our system as safe as possible... like, defend against the outside intruder...like, a strong security culture among our staff... try to make ourselves as safe as possible... So that is more standalone basis... a "do it ourselves" attitude (Personal Communication, 26 March 2006).

Another respondent, a regional director for security of a global credit card company interviewed in early 2006 also said that bank/police liaison was not always as close as it seems:

> I can say this is very rare. The private sector people attend the regular meetings with the police discussing fraud cases ... Certainly there are

some exceptional cases, where they need to receive assistance (Personal Communication, 3 February 2006).

Opinion thus seems mixed as to whether the police and the banks work closely together. However, the case studies do provide evidence that the police and the banks have a close working relationship. For example, in Cases 12 and 15, the prosecution called employees from the HSBC as expert witnesses. This kind of liaison confirms the link between the criminal justice system and the banks, and the finding that the bank staff provide forensic evidence for the prosecution. Case 11 also showed that the HKMA cooperated closely with the SFC to monitor and refer a case for prosecution. The manager of Internet banking at the HSBC interviewed in early 2006 also confirmed that she and her organization regularly dialog with the police on matters relating to Internet security. However, she also added that the discussions were not face-to-face talks with the police, because "[w]e have a dedicated unit to do that... we cooperate because it is necessary...We have that two-way channel." She went on to explain that "our security division is made up of ex-policemen... we work with the police, we like to maintain a close working relation with them..." (Personal Communication, 3 March 2006).

Conclusions

On the issue of whether the banks and the police closely liaise with one another in investigating and prosecuting Internet banking fraud, the evidence from interviews is more mixed than that from the cases examined, which seem to show a close level of liaison. There is also strong evidence from the sample cases that the police are capable of successfully investigating Internet banking fraud, even when the crimes are trans-jurisdictional. However, the police may have some problems when the cases involve mainland Communist Party officials, as these "official" figures tend to be beyond the reach of Hong Kong police, and law may not be able to reach them.

The cases examined also suggest that the police use the Computer Crimes Ordinance 1993 as a "catch-all" to initiate investigations but then change the charges to traditional crimes at trial. There is also clear evidence from the cases to suggest that the police prosecute minor cases (i.e., trivial, low monetary value crimes) and cases involving "big fish" (i.e., serious, high monetary value crimes).

Furthermore, there is some evidence that mainlandization is influencing Internet banking crime in Hong Kong, as many of the cases involve mainland parties and parties traveling to mainland China to do business. Some of the cases involved corruption, while others used the Internet in mainland China to perpetuate crimes in Hong Kong. As these cases suggest that mainland practices are seeping into Hong Kong, these types of crimes may give rise to the fear that Hong Kong's reputation as an international financial center could be damaged if such practices go unchecked.

Finally, the high number of appeal cases in the sample suggests that the ideology of the rule of law is alive and well in Hong Kong. However, the high number of custodial sentences imposed on defendants in the sample and the prosecution of minor crimes also suggest that the Hong Kong police use the law for purposes of crime control rather than due process.

Bibliography

Bloomberg Business Week (2003) "The Bank of China's Real Scandal," 19 June, http://www.businessweek.com/stories/2003-06-19/the-bank-of-chinas-real-scandal (accessed 27 June 2013).

Deloitte Touche Tohmatsu (2012) *Report on Statistics on Mainland IPOs*, http://www.deloitte.com/view/en_CN/cn/Pressroom/pr/98b4f30f818fb310VgnVCM3000003456f70aRCRD.htm (accessed 11 April 2013).

Linklaters (2010) *Regulatory Investigations Update*, 20 November, http://www.linklaters.com/pdfs/mkt/lit/Regulatory_Investigations_Update_November2010.pdf (accessed 27 May 2013).

Lam Siu Po v Commissioner of Police (2009) Court of Final Appeal HKCFA 24.

Wing Lo, T. and Ngan, Paul (2009) "Restricting Loans of Money to Hong Kong Civil Servants: Social Censure or Violation of Human Rights?," *Crime, Law, and Social Change* 52: 385–403.

Case List

Case 1: HKSAR v Tsun Shui Lun-[1999] 2 HKC 547, Magistracy Appeal No 723 of 1998

Case 2: HKSAR v Tam Hei Lun & ORS-[2000] 3 HKC 745, Magistracy Appeal No 385 of 2000

Case 3: HKSAR v Choy Yau Pun-[2002] 4 HKC 309, Magistracy Appeal No 450 of 2002

Case 4: HKSAR v Lai Mei Yuk [2004], Criminal Appeal No 427 of 2003

Case 5: RE Chen Kam Chiu and Others-[2005] HKCU 987, CACC 179/2004 (On Appeal HCCC No 158 of 2003)

Case 6: Chiu Hoi Po v Commissioner of Police-[2006] HKCU 681, HCAL 105/2003

Case 7: HKSAR v Leong Wai Keong-[2008] HKCU 1915, Court of Appeal CACC 476/2007

Case 8: HKSAR v Kam To Fung-[2008] HKEC 1041, Magistracy Appeal No 565 of 2007

Case 9: HKSAR v Marimuthu Jaisanka-[2008] HKCU 243, Court of Appeal CACC 403/2007

Case 10: HKSAR v Ho Ching Wah-[2009], Court of Appeal CACC 106/2008
Case 11: HKSAR v Ma Hon Kit Sammy & ORS-[2010] HKCU 2189, Court of Appeal
       CACC 148/2009
Case 12: HKSAR v Cai Zhaorong-[2012], Court of Appeal CACC 365/2011
Case 13: HKSAR v Chan Wai Ming-[2012] DCCC 137B/2011
Case 14: HKSAR v Cheung Yiu Ming-[2012] DCCC 320/2011
Case 15: HKSAR v Yaghi Jose-[2012] DCCC 141/2012

7

The Role of Spam in Cybercrime: Data from the Australian Cybercrime Pilot Observatory

Mamoun Alazab and Roderic Broadhurst

Introduction

The use of the Internet for the purpose of crime is a rapidly growing phenomenon that requires a proactive and coordinated response (UNODC 2013). Cybercriminals increasingly use sophisticated tools and methods to distribute a wide range of malicious content, often combining deceptive "social engineering" tricks with spam emails, hosting of phishing sites, and identity theft (Smith and Hutchings 2014). Unsolicited bulk, mass emails, or "spam," pose a global challenge because this remains a major vector for the dissemination of malware.

Spam takes many forms, and many varieties have been described in the European Commission study (European Commission 2009). Spam can merely carry annoying but benign advertising; however, it can also be the initial contact point for cybercriminals, such as the operators of a fraudulent scheme, who use emails to contact and solicit prospective victims for money (as in advance fee frauds) or to commit identity theft by deceiving recipients of such mail into sharing personal, bank, and financial account information.

Unlike "low volume, high value" cybercrime that targets financial services and requires advanced hacking capability, spam enables malware to reach "high volume, low value" targets that are less likely to have effective anti-virus or other countermeasures in place. Such malware is distributed through two types of spam: those with an attachment that contains a virus or Trojan that installs itself in the victim's computer when the attachment is opened and those with a hyperlink to a web page where the malware is downloaded onto the compromised computer (Alazab and Venkatraman 2013).

Spam emails with hidden malware or URLs that direct users to malware are common methods used by cybercriminals to find new victims. For example, spammers may want to expand their botnets, or cybercriminals may use them to propagate their computer intrusion software (i.e., software developed as "crime-ware") so as to harvest passwords, credit card accounts, bank accounts, and other sensitive personal information. Our research aims to develop preventative methods to help reduce the propagation of malware via the frequently used medium of spam emails. Before presenting our results, we briefly describe our data and how criminals disseminate spam emails.

Spam thrives on the acquisition of active email addresses, and these addresses are harvested in three different ways: first, by searching for email addresses listed on websites and message boards; secondly, by performing a "dictionary attack," a combination of randomly generated usernames with known domain names to guess correct addresses; and finally, by purchasing address lists from other individuals or organizations such as in underground markets (Takahashi et al. 2010). In 2012, Trend Micro reported that more than half of the total number of spam email addresses collected from February to September 2012 were obtainable from websites alone (Trend Micro 2012). Once email addresses are harvested, spammers distribute spam by using botnets, and this technique is used by large spam botnets, like Storm Worm, Grum, Bobax, Cutwail, Mega-D, Maazben, Pushdo, Rustock, and others (Stringhini et al. 2011).

The proliferation of botnets has enabled large quantities of spam emails to be sent in a coordinated way, amplifying cybercrime activities (Alazab et al. 2013). Botnets are networks of compromised computers controlled by a "botmaster" who often rents the botnet to spammers and others that use them to deliver malware. Botnets are the backbone of spam delivery, and estimates suggest that about 85 percent of the world's spam email were sent by botnets (John et al. 2009) (Symantec 2010) (McAfee 2010) (Stringhini et al. 2012). Greater volumes of spam increase the chances of an effective delivery of attachments and/or URLs that include malware. Spam is an important vector for spreading malware and along with better social engineering has enhanced the ability to deceive many into malware self-infection.

Botnets sending large quantities of spam emails require a C&C or command-and-control botnet to coordinate targets and evade blacklisting services. The use of botnets shows how spammers have learned to manipulate the networks of compromised computers and servers around the world to ensure high volumes of spam are delivered. Botnet-based spam appears to have emerged around 2004 as a type of novel advanced distribution network (also associated with DDoS attacks) and responsible for almost all large-scale spam campaigns.

Fully automated spam campaigns often include malicious spam created by crime-ware toolkits, such as Blackhole, and as noted can insert via malicious URLs or malicious attachments via advanced intrusion software. Spam often contains a malicious attachment or a link to legitimate websites that have been compromised by a web attack toolkit. These toolkits are easy to use and efficiently leverage existing vulnerabilities. For instance, the well-known Blackhole attack toolkit has been applied in some spam campaigns. A recent criminal innovation involves attacking computers indirectly by concealing intrusions through an intermediary website or "waterhole," sites that the target is likely to visit and that also host malicious code on the landing page (see Figure 7.1). Cybercriminals also create links in spam messages that point to exploit portals hosting Blackhole (or similar malware), an alternative approach that avoids the need to hack legitimate websites before planting malicious code.

Automation occurs using a template format for spam content distributed through thousands of compromised computer hosts (or "zombies") available to a botmaster who charges fees to do so. Consequently spam-driven malicious attacks, often botnet applications, have become more organized and targeted. Compromised computers are infected with software bots that allow the "zombie" computers to be controlled remotely

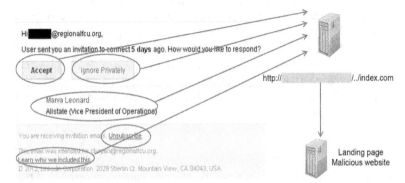

Figure 7.1   Example of redirect link 'Waterhole' attacks

through an established command-and-control channel (C&C). Our analysis shows that 40 percent of our data set consist of emails that have been distributed more than 50 times and sometimes more than 1,000 times (either with the same or different attachments), suggesting that these spam emails have been sent by different groups, using botnets to distribute them.

Volume

Beside its potential for crime, spam is problematic because of its sheer volume, which impedes the flow of legitimate Internet traffic around the world. In 2010, security service provider Symantec estimated that spam accounted for more than 90 percent of the total global email traffic (Symantec 2010). Estimates have subsequently been lowered to an average of 66.5 percent of all emails sent in the first quarter of 2013, and on average 3.3 percent of these mails contained malicious attachments (Gudkova 2013). Cisco reported that spam volumes had grown with global volumes now reaching over 200 billion messages per month (Cisco 2014), but Symantec estimated higher volumes of about 30 million spam emails each day (Symantec 2013). Our data sets show that spam, depending on the source, contains variable levels of suspect mail. Although levels of malicious spam may seem insignificant at the individual level, it is estimated that in 2013, approximately 183 billion emails were sent and received every day, and so the number of malicious communication can be substantial (Radicati and Levenstein 2013). It is not surprising that a huge amount of spam emails are necessary because MessageLabs estimated that for a spam advertisement to be profitable, 1 in 25,000 spam recipients needed to open the email and make a purchase in an underground market (Symantec 2008). Spam sent in 2010 earned its operators USD$2.7 million in profit from fake sales in pharmaceuticals alone (Krebs 2012), while the cost of spam to Internet service providers (ISPs) and users worldwide reaches the billions (Anderson et al. 2013). A recent study on the economics of spam (Rao and Reiley 2012) calculated that spammers may collect gross global revenues of the order of USD$200 million per year, while some $20 billion is spent fending off unwanted emails.

Data set

We use three real-world data sets (DS) of spam emails collected in 2012. Emails are identified as spam in two ways: first, an email user may determine that an email is spam; secondly, emails may be collected and identified as originating from known spamming networks. Both scenarios are captured in our real-world data sets.

The first data set, the HABUL DS, comes from the HABUL plugin for Thunderbird (an open-source cross-platform email service and usenet based on Mozilla code) that uses an adaptive filter to learn (machine learn variants of spam text) from a user's labeling of emails as spam or normal email. The second data set, is an automated collection from a global system of honeypots and spam-traps designed to monitor information about spam and other malicious activities, which we labeled the Botnet DS. The third data set are suspected spam emails reported by Australians and sourced from the Spam Intelligence Data-Set (SID, provided by the Australian Communication and Media Authority), which we labeled the "OzSpam" DS. Table 7.1 summarizes the descriptive statistics of the three DS feeds. The HABUL DS is much smaller than the Botnet DS and OzSpam DS but has the advantage that the emails collected have been manually labeled as spam by recipients. While the spam in the HABUL DS has been viewed by a potential victim, the Botnet DS and OzSpam DS contain spam that circulated all over the world but without the certainty that the emails have reached their intended targets. All together about 13.5 million spam emails were collected, which included nearly half a million attachments and over six million URLs. The number of emails collected from these three sources is reported in Table 7.1 by month of delivery. The proportion of spam that carried malicious code in attachments or through hyperlinks in the body text of the email varied across the different sources. For example, 1.38 percent of HABUL attachments and 13 percent of HABUL hyperlinks where identified as malware, and this was similar to the OzSpam DS, which identified 10.5 percent of hyperlinks and 0.77 percent of attachments as malware. The Botnet DS, however, had fewer suspect hyperlinks (0.52%) than expected given the method applied but approximately similar proportions of malware in the attachments (0.95%) forwarded with the spam mail.

The three data sets also showed some general similarities: between 39 percent of HABUL DS and 46.3 percent of OzSpam DS spam emails contained at least one URL, the Botnet DS included 44.9 percent of emails with a URL, of which only 1.15 percent of the links contained malware. Interestingly almost one-quarter (22.7%) of OzSpam DS URLs were identified as malicious, and this was much higher than found in HABUL (14.6%) and the 1.15 percent identified in the Botnet DS. In contrast, higher proportions of emails that include an attachment were malicious in HABUL (34.5%) and Botnet DS (50%) but were similar to levels of web-link malware found in the OzSpam DS (21%). For each data set, there were peak periods of spam that either contained

Table 7.1   Data set statistics

1- HABUL Data Set

Month	Emails	With Attachments		With URLs	
		Total	Mal.	Total	Mal.
January	67	7	3	25	3
February	104	10	2	33	6
March	75	5	0	28	4
April	65	4	2	26	2
May	83	4	0	38	5
June	94	1	0	41	5
July	72	2	1	26	11
August	85	0	0	46	10
September	363	11	7	140	4
October	73	1	1	11	3
November	193	4	0	89	13
December	95	6	3	31	12
Total	1,369	55	19	534	78

2- Botnet Data Set

Month	Emails	With Attachments		With URLs	
		Total	Mal.	Total	Mal.
January	31,991	139	27	12,480	4
February	49,085	528	66	14,748	4
March	45,413	540	52	19,895	23
April	33,311	328	175	12,339	0
May	28,415	753	592	13,645	3
June	11,587	102	56	8,052	80
July	16,251	425	196	5,615	92
August	21,970	291	113	16,970	707
September	27,819	282	12	17,924	442
October	13,426	899	524	4,949	2
November	17,145	1,107	882	7,877	49
December	20,696	621	313	7,992	241
Total	317,109	6,015	3,008	142,486	1,647

3- OzSpam Data Set

Month	Emails	With Attachments		With URLs	
		Total	Mal.	Total	Mal.
January	17,512	942	99	6,793	1,398
February	867,725	86,466	38,556	381,304	83,327
March	297,999	19,306	4,769	129,051	27,511
April	1,026,077	62,014	17,406	405,475	75,801
May	1,084,214	59,717	14,675	430,242	87,699
June	1,521,541	32,663	5,136	594,092	104,080
July	1,710,530	36,555	3,859	744,368	260,084
August	1,102,896	42,764	3,724	552,584	104,476
September	1,069,399	47,744	4,145	539,869	80,276
October	937,430	42,966	4,312	418,754	68,399
November	1,194,951	34,164	3,175	674,578	136,686
December	2,301,783	21,607	2,382	1,210,647	356,225
Total	13,132,057	486,908	102,238	6,087,757	1,385,962

malicious content or not and which suggested different types of spam (mass propagation) campaigns. These campaigns usually shared similarities in the content of their emails, and this alone may indicate the risk of malicious content. These variations in the prevalence of malware nevertheless show the current importance of hyperlinks embedded in email spam as an important vector for compromising a computer and inserting malware.

We received the data sets in anonymized form, so no identifiable email addresses or IPs are available for analysis. Future work will include, where possible, metadata of this kind. Our analysis only looked at the spam, which appeared to have been relayed through Australia, for example the last hop IP address was in Australia. Our spam data sets are composed entirely of emails that had already breached existing filters, both at the provider and end-user level.

We limit our analysis to malicious spam, which contains at least one malicious attachment or URL. Therefore, for each email, we extracted attachments and URLs and uploaded them to VirusTotal, an online virus checker that offers support for researchers, and scanned for viruses and suspicious content. VirusTotal uses over 40 different virus scanners, and we considered an attachment or URL to be malicious if at least one scanner showed a positive result.

Social engineering and spear phishing

A common way to convince recipients of spam emails to act on these emails is the use of "social engineering" techniques that deceive the recipients into believing that the email is legitimate (Broadhurst and Chang 2013). Cybercriminals favor social engineering tactics to persuade their victims to click on a malicious URL or download malware because it is easier than trying to insert malware remotely (such as Trojans) so that key-loggers can obtain banking passwords or other sensitive information. Social engineering involves the use of deception to manipulate behavior by building rapport with the victim then exploiting that emerging relationship to obtain information or access to the computer (Chantler and Broadhurst 2006). For example, spam emails often use alarming language (e.g., "your bank account has been suspended") to convince users to click on the malicious URL and rectify the problem (FireEye 2013). The words used within these malicious file name attachments provide some insights into the tactics that cybercriminals use. Often these names are related to trusted brands, for example "FedEx" or "PayPal," that act as lures. In examining the malicious attachments file names in our data, we found a trend to reference trusted business labels rather than other labels or general names. Our results show labels or brands related to shipping are among the most common (e.g., FedEx, UPS, USPS, and DHL). Figure 7.2 shows how benign file names and extensions are often used by criminals and can also pose a security threat.

Spear phishing is a spamming method that targets selected users or groups via a compromised computer that can then be used as a "slave" or "zombie" computer capable of importing malware (key-loggers, crypters, etc.) to steal banking passwords and other confidential data. Spear-phishing emails are personalized and often try to impersonate a trusted source to avoid anti-spam detection at the system level. Our analysis showed that spammers use common business terms in their file names as spear phishing enticements as noted in the adoption of common logistics brands. The most commonly found shared file types using purloined brands had the file extension .zip and were responsible for 76 percent of the total number of spear-phishing emails attachments during our monitoring period (see Figure 7.2). File extensions bearing other common formats such as .pdf, .xls, .doc, .jpg, .txt, .bmp, and .gif accounted for the remaining 24 percent of malware-infected formats, and of these .jpg and .txt extensions accounted for most of the remainder. Spear phishing is an effective cyber-attack technique and the

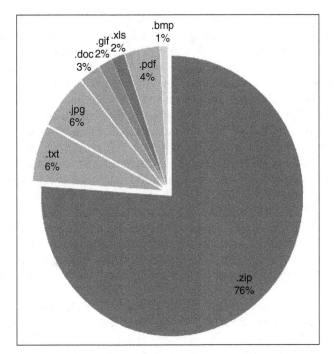

Figure 7.2   Top ten malicious file extensions

most common method for initiating advanced malware campaigns. The number of recipients was usually one victim for the majority (69%) of emails, and the number of attachments was also limited to one rather than multiple or bundled files in 88 percent of emails. This suggests that spammers have learned to focus only on sending a single malicious attachment and craft the payload necessary to get that attachment to the end-user. It is likely that we will see more than one attachment in future spam runs and continued variation in subject lines (e.g., "World Cup," "Missing Malaysia Aircraft") as spammers harvest headlines from news websites and blogs (Symantec 2010) in the constant search for new ways to deliver malicious attachments to end-users.

Compressed files

Spam emails can carry different types of files as attachments; however, it appears that files disguised under the extension (.zip) are the most common malware file type. The majority of spam solutions block email attachments with the (.exe) file extension but do not reliably scan archived and zipped documents (Krebs 2011), therefore encouraging spammers to compress executable files (.exe) into an archived form such as .zip, .rar, and .tar. Our analysis showed that .zip files represent the vast majority (90%) of malicious files. The .zip attachments can contain many different file types, and because users remain unaware of the risks with these file extensions, they are effective for disseminating malware.

With the increase in the rate of malicious attachments carried by spam, security firms and other organizations often block or reject emails that contain, as noted above, an attachment with an executable file (e.g., with the extension .exe), prompting criminals to use a double extension to disguise the executable file. Example: tracking_instructions.pdf.zip, the first extension is that of a benign attachment (e.g., in this example .pdf but might be a .jpg or some other common file), but the second extension represents what the file really is – a .zip file containing .exe files; a gap between the two extensions sometimes can prevent simple spam filters from determining that the attachment is really an executable file.

There are malware formats that also try to get a recipient to download them using the double extensions method (e.g., as per .doc.exe). For example, Kraken Botnet attempted to avoid .exe blocking filters by sending executable files in a compressed form (like .zip). Other detection avoidance measures use double extensions (.jpg.exe) to try to trick users or filters. Recipients will see .jpg or .pdf and may feel comfortable to open up what seems to be an image or a standard pdf file. Our analysis of the three data sets confirms that these simple avoidance methods are still commonplace.

Right-to-left override email attacks

The right-to-left override (RLO) character refers to a special character within Unicode (U+202e), which in turn is an encoding system that enables computers to exchange information regardless of the language used. It supports languages that are written right-to-left, such as Arabic and Hebrew. RLO can be used to disguise a malicious file (Krebs 2011).

Usually when executable files are decompressed their appearance provokes suspicion. So spammers conceal executable files with fake icons to make them appear as harmless file extensions, such as, .pdf, .doc, .xls, or .jpg) and employ the Unicode's RLO technique, which reverses the character ordering from right to left and so changes the order of characters.

Spammers thus use the RLO technique to deceive users into downloading and executing the malicious file hidden in an attachment under the cover of a fake filename extension, and the technique is often combined with very long file names which are also employed to disguise the .exe file name extension. Although this deception has been known for some time (Davis and Suignard 2013), it can be hard to detect; an example of how it works is given in Figure 7.3. To make the process even harder, the malicious file names manipulated in this method are also delivered within zip files or archives. The example in Figure 7.3 shows how an executable file can be sent disguised as a .doc format using the RLO technique: the attachment name appears to look like a Word document file; however, the real file is an executable file (.exe) with the RLO reverse command activated by placing the Unicode command (U+202e) for right-to-left override just before the "d" in "doc" in the actual attachment name.

Figure 7.4 shows the breakdown of the RLO technique, and it shows the malicious contents are mostly masked as Adobe Acrobat portable

---

The actual attachment name is:
Your_personal_details_15.6.2012_E.**heydoc.exe**

By placing the right to left override command just before the "d" in "doc"
As a result the new file name will appear like a word document:
Your_personal_details_15.6.2012_E.**heyexe.doc**

Figure 7.3   Example of a 'Right-to-Left Override' email attack

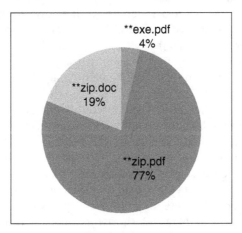

Figure 7.4   Statistics on the use of RLO email attacks

document files (pdf) and Microsoft Office Word documents (doc). It seems that pdf and doc provide spammers with an effective means of disguising attacks because users trust them but lack awareness that malware can be disguised as these file types.

URL shortening

A service called "URL shortening" has become popular especially among social network users and promoters, but it also enables a variant of methods used to disguise or obfuscate spam/malware. This service allows long URLs to be transformed into much shorter URLs and thus enhances the likely use of the link. Spammers also use URL-shortening services, even establishing their own shortening services (Symantec 2012). Inadvertently, this "shortening" allows spammers to deceive recipients and hide the true destination/location of the link and also avoid detection. Also, some URL-shortening services provide feedback through a dashboard that shows the number of clicks, locations, and Internet browsers. This added-value feature enables spammers to find out what links engage users and thus resend tailored links and so improve the effectiveness of their campaigns.

Spammers ideally redirect a link through many different shortened links; rather than leading straight to the spammer's final destination website, the links point to a shortened URL on the spammer's fake URL-shortening website before redirecting to the spammer's final website and its hidden malicious content. This service has become more common because of its simplicity, automated capability, and anonymity.

There are many URL-shortening services that spammers can use or create. Popular URL shortener websites such as Google URL Shortener and Bit.ly provide an easy interface that allows users to convert long URLs into short URLs. After receiving a long URL, the services use a hash function to map the long URL to a short string of alphanumeric characters, which is then added to the domain name of the shortener and returned as the short URL.

Symantec and other information security companies have warned that attacks using URL-shortening services are likely. These attacks might occur because a criminal group gains control of a legitimate URL-shortening service or creates such a service which appears and operates legitimately (often hosted in Russia and the Ukraine, or with Russian domain names), before being turned to malicious use (Symantec 2012). Our analysis confirms these findings (Wang et al. 2013), and Table 7.2 shows the use of legitimate URL-shortening services in spam emails in

Table 7.2 URL-shortening service used

Name	URL-shortening Service Name	% of use
Twitter	t.co	56.3
Google	goo.gl	29.9
HootSuite	ow.ly	3.5
ViralURL	vur.me	3.4
Bitly	j.mp	3.0
TinyURL	tinyurl.com	0.7
ViralURL	vurl.bz	0.6
is.gd	is.gd	0.6
scrnchlet	scrnch.me	0.5
Bitly	bitly.com	0.5

all of our DS. The table shows that URLs shortened via Twitter accounted for 56 percent of these events. Due to its large audience and reach, spammers use Twitter's URL shortener to entice users because they look like a tweet.

Discussion

Cybercrime has rapidly evolved, and attacks via email (spam) remain one of the major vectors for the dissemination of malware and many predicate forms of cybercrime. Spam as prime means for "social engineering" continues to be one of the popular ways to spread and inject malware on computers and other digital devices. Unlike cybercrime that targets (low volume, high value) victims such as banks and financial services that often requires advanced hacking capability, spam enables malware to reach "high volume, low value" targets that are less likely to have effective anti-virus or other countermeasures in place. The deceptions achieved through "social engineering" of email messages are usually well known, but less is known about whether different forms of social engineering are related to different types of malware and cybercrime. Malware is often distributed in one of two types of spam messages: with an attachment (e.g., Trojan) or with a link to a web page or a hyperlink that enables malware to be downloaded onto the compromised computer.

Existing detection and defense mechanisms to deal with email spam containing malicious code are mostly reactive and ineffective against constantly novel spam email formats that hide ever-improving malware payloads and capabilities. There is an urgent need to identify new malware-embedded spam attacks (either in the increasingly common URL approach or via disguised attachments) as rapidly as possible without the need to wait for updates from spam scanners or blacklists. These scanners attempt to filter suspect emails but often cannot update as quickly as the novel variants of malware and/or social engineering appear and necessarily rely on many different resources to identify malicious content (Tran et al. 2013).

One popular countermeasure against this type of malicious spam email is to develop machine learning techniques that routinely check on the content of the email, attachments, and suspect URLs and can match suspect websites or URLs, attachments, and socially engineered email content with known "blacklists" of such activities. Machine learning methods of identifying spam and other spam filtering methods aim to be highly responsive to changes in spamming techniques and often reveal the scope and nature of the malicious content but thus far have not been sufficiently flexible to handle variations or permutations in the content or delivery methods found in spam emails (Blanzieri and Bryl 2008).

Technical measures or government interventions alone cannot tackle spam, and it is important that law enforcement agencies (LEAs) have the capacity to effectively identify, investigate, and prosecute cybercriminals who use spam as a means of delivering malware. A key solution is the creation and maintenance of effective partnerships between LEAs, regulatory agencies, and stakeholders such as ISPs. Support from the private sector is vital. Given that with few exceptions most spam-malware originates outside Australian cyberspace, the challenge of creating an effective international legal response for the suppression of cybercrime remains a key if long-term goal. Engagement with international (e.g., Council of Europe's Convention on Cybercrime) and regional efforts to suppress cybercrime, combined with the active involvement of industry and civil society, continue to offer the best means for effective suppression of the malware-spam problem.

Attention to high-level cybersecurity issues such as the protection of national critical infrastructure is a clear concern, but vulnerabilities are broadly spread throughout the online community. Household users and small enterprises are often vulnerable to cyber attack due to many factors including the cost of up-to-date security. Thus the oft-repeated cliché that our security is only as good as our weakest link applies. Confidence in the safety of the digital world and in the performance of e-commerce and e-government are predicated on low risks. The impact of "high volume, low value" cybercrime undermines this confidence.

Indeed a highly effective and secure e-commerce environment is vital and can offer a very attractive market for economies and business vulnerable to cybercrime in their own jurisdictions.

In short, fighting spam requires a combination of technology and relevant up-to-date laws and policies as well as the constant reformulation of crime prevention practices to keep abreast of the evolution of spam-malware techniques. Shifts in malware attacks to new vectors using spam-like methods often based on astute and tailored social engineering also need constant attention. A good example is the shift to Twitter and other new media where, for example, URL shortening methods may prosper. While effective civil measures (including anti-address harvesting laws) are in place to mitigate commercial misuse of spam in Australian cyberspace, the challenge lies in the integration of countermeasures that can further suppress the spam-malware vector. In addition, maintaining the high-level industry-government-LEA coordination required for successful disruption of malware-driven spam campaigns and other cybercrime must be at the forefront of government-led initiatives. To maximize such cooperation, a reassessment of the co-regulatory burdens on industry may be required and proposals to deregulate the current e-marketing and spam industry codes of practice, for example, may be welcome if they encourage more self-help and help secure continued partnership with government in the fight against cybercrime.

Consumer education and general high awareness of the risks of "self-inflicted" harm associated with malware-driven spam and Internet scams and attacks has to be a priority for small enterprises and household users – in short, the "weakest links" must also receive the most timely prevention advice possible. Now more than ever there is a need for more extensive and well-funded crime prevention programs, on the scale normally associated with major public health awareness campaigns that increase self-help across the online community. Proactive predictive technical measures should also continue to be a key focus in the fight against spam and may provide a potentially effective vehicle to degrade the volume of spam-malware and the risks associated with high volume, low value cybercrime.

In spam emails "crimeware-as-a-service" is more evident than ever, and it involves selling exploits and tools for computer trespass. Once attack tools are in place, buyers can rent them to deliver a variety of attacks. Individuals and businesses need to increase their awareness of the dangers of spam emails, especially targeted attacks (e.g., spear phishing) and create effective policies and practices to prevent its distribution.

Botnets generally account for the global dissemination of spam. The widespread use of botnets shows how spammers manipulate the networks of infected computers and servers around the world to ensure high volumes of spam are delivered. The increased role for networked crime groups has also impacted on the scale and sophistication of cybercrime. Trends in spam designed to create botnets also show increasing malware complexity that exploits new opportunities arising from automated financial activities (e.g., *GameOver ZeuS* and *CryptoLocker*). The Internet has also become the preferred platform to deploy spam attacks to intentionally disrupt or to subvert these automated services and also to launch DDoS attacks (for profit or ideological motives). These tools have far-reaching implications for the evolution of cybercrime and need to be explored and investigated.

Spear phishing is a good example of offender innovation. By using personal information already gathered through deception or by inserting a remote access tool (RAT) via an email with apparently *relevant* content from a trusted sender, this method can circumvent countermeasures. So the rapid emergence of bespoke email content tailored to entice a specific victim or victim type via "spear phishing" also pose dangers that require equally targeted education and crime prevention efforts. The use of spam emails remains an important and underestimated vector for the propagation of malware. The forms of social engineering used in spam emails have also become more sophisticated, personalized, and compelling, thus improving the ability to deceive many users into malware self-infection. Constant education and the development of crime prevention practices that focus on methods of deception are crucial and need to be as current as the novel and advanced forms of malware present (e.g., SCAMwatch, www.scamwatch.gov.au/; and ACMA's Cybersmart http://www.cybersmart.gov.au/).

Acknowledgments

We acknowledge the assistance of the Criminology Research Council (Grant CRG 13/12–13) and the Australian Research Council. We also thank the Australian Communications and Media Authority (ACMA) and the Computer Emergency Response Team (CERT) Australia for their assistance in the provision of data and support. We thank our colleagues Peter Grabosky, Khoi-Nguyen Tran, Ki-hong "Steve" Chon, and Brigitte Bouhours for their contributions to the data collection and analysis.

Bibliography

Alazab, M., Layton, R., Broadhurst, R. and Bouhours, B. (2013) "Malicious Spam Emails Developments and Authorship Attribution, IEEE," in *The Fourth Cybercrime and Trustworthy Computing Workshop* 58–68, Sydney: IEEE.

Alazab, M. and Venkatraman, S. (2013) "Detecting Malicious Behaviour Using Supervised Learning Algorithms of the Function Calls," *International Journal of Electronic Security and Digital Forensics* 5(2): 90–109.

Anderson, R., Barton, C., Böhme, R., Clayton, R., Eeten, M., Levi, M., Moore, T. and Savage, S. (2013) "Measuring the Cost of Cybercrime," in *The Economics of Information Security and Privacy*, R. Böhme (ed.), Berlin Heidelberg IV: Springer, 265–300.

Blanzieri, E. and Bryl, A. (2008) "A Survey of Learning-based Techniques of Email Spam Filtering," *Artificial Intelligence Review* 29: 63–92.

Broadhurst, R. and Chang, L. (2013) "Cybercrime in Asia: Trends and Challenges," in *Handbook of Asian Criminology*, J. Liu, B. Hebenton and S. Jou (eds.), New York: Springer, 49–63.

Chantler, A. and Broadhurst, R. (2006) *Social Engineering and Crime Prevention in Cyberspace*, Technical Report, Justice, Queensland University of Technology, http://eprints.qut.edu.au/7526/1/7526.pdf (accessed 1 October 2014).

Cisco (2014) *Spam Hits Three-Year High-Water Mark*, http://blogs.cisco.com/security/spam-hits-three-year-high-water-mark/ (accessed 1 October 2014).

Davis, M. and Suignard, M. (2013) *Unicode Technical Report #36 "Unicode Security Considerations,"* http://unicode.org/reports/tr36/#Bidirectional_Text_Spoofing (accessed 1 October 2014).

European Commission (2009) *EU Study on the Legal Analysis of a Single Market for the Information Society: New Rules for a New Age?*, Digtial Agenda for Europe, http://ec.europa.eu/information_society/newsroom/cf/itemdetail.cfm?item_id=7022& (accessed 1 October 2014).

FireEye (2013) *FireEye Advanced Threat Report – 2H 2012*, FireEye Inc., http://www2.fireeye.com/rs/fireye/images/fireeye-advanced-threat-report-2h2012.pdf (accessed 1 October 2014).

Gudkova, D. (2013) *Spam in Q1 2013*, Securelist, http://www.securelist.com/en/analysis/204792291/Spam%20_in_Q1_2013 (accessed 1 October 2014).

John, J., Moshchuk, A., Gribble, S. and Krishnamurthy, A. (2009) "Studying Spamming Botnets Using Botlab," in *The 6th USENIX Symposium on Networked Systems Design and Implementation*, Boston: 291–306.

Krebs, B. (2012) *Who's Behind the World's Largest Spam Botnet?*, Krebsonsecurity, http://krebsonsecurity.com/2012/02/whos-behind-the-worlds-largest-spam-botnet/ (accessed 1 October 2014).

Krebs, B. (2011) *"Right-to-Left Override" Aids Email Attacks*, http://krebsonsecurity.com/2011/09/right-to-left-override-aids-email-attacks/ (accessed 1 October 2014).

Radicati, S. and Levenstein, J. (2013) *Email Statistics Report, 2013–2017*, The Radicati Group Inc., http://www.radicati.com/wp/wp-content/uploads/2013/04/Email-Statistics-Report-2013-2017-Executive-Summary.pdf (accessed 1 October 2014).

Rao, J. and Reiley, D. (2012) "The Economics of Spam," *Journal of Economic Perspectives* 26(3): 87–110.

Smith, R. and Hutchings, A. (2014) "Identity Crime and Misuse in Australia: Results of the 2013 Online Survey," *Research and Public Policy Series,* Canberra, Australia: Australian Institute of Criminology, http://aic.gov.au/media_library/publications/rpp/128/rpp128.pdf (accessed 1 October 2014).

Stringhini, G., Egele, M., Kruegel, C. and Vigna, G. (2012) "Breaking the Loop: Leveraging Botnet Feedback for Spam Mitigation," in *The Seventh Annual Graduate Student Workshop on Computing,* Santa Barbara, CA: University of California, 25–6.

Stringhini, G., Holz, T., Stone-Gross, B., Kruegel, C. and Vigna, G. (2011) "BOT-MAGNIFIER: Locating Spambots on the Internet," in *The 20th USENIX Conference on Security,* San Francisco, CA: USENIX Association, 1–16.

Symantec (2013) *Internet Security Threat Report 2013* vol. 18, <https://scm.symantec.com/resources/istr18_en.pdf>.

Symantec (2012) *Symantec Internet Security Threat Report: Trends for 2011* vol. 17, http://www.symantec.com/content/en/us/enterprise/other_resources/b-istr_main_report_2011_21239364.en-us.pdf (accessed 1 October 2014).

Symantec (2010) *MessageLabs Intelligence: 2010 Annual Security Report,* https://www.inteco.es/file/27gHxrzWsYyeyRTFYq8MuQ (accessed 1 October 2014).

Symantec (2008) *MessageLabs Intelligence: 2008 Annual Security Report,* http://www.ifap.ru/pr/2008/n081208a.pdf (accessed 1 October 2014).

Takahashi, K., Sakai, A. and Sakurai, K. (2010) "Spam Mail Blocking in Mailing Lists," in *Multimedia,* K. Nishi (ed.), InTech.

Tran, K.-N., Alazab, M. and Broadhurst, R. (2013) "Towards a Feature Rich Model for Predicting Spam Emails Containing Malicious Attachments and URLs," in *Eleventh Australasian Data Mining Conference,* P. Christen, P. Kennedy, L. Liu, K.-L. Ong, A. Stranieri and Y. Zhao (eds.), Canberra ACT vol. 146, Canberra, Australia: Australian Computer Society.

Trend Micro (2012) *Spear-Phishing Email: Most Favored APT Attack Bait,* Trend Micro, http://www.trendmicro.com.au/cloud-content/us/pdfs/security-intelligence/white-papers/wp-spear-phishing-email-most-favored-apt-attack-bait.pdf (accessed 1 October 2014).

UNODC (2013) *Comprehensive Study on Cybercrime,* http://www.unodc.org/documents/organized-crime/UNODC_CCPCJ_EG.4_2013/CYBERCRIME_STUDY_210213.pdf (accessed 1 October 2014).

Wang, D., Navathe, S., Liu, L., Irani, D., Tamersoy, A. and Pu, C. (2013) "Click Traffic Analysis of Short URL Spam on Twitter," in *The 9th IEEE International Conference on Collaborative Computing: Networking, Applications and Worksharing,* Austin: IEEE, 250–9.

8
Quantifying Sexually Explicit Language

George R. S. Weir

Acknowledgments

I am grateful for practical insights from several former students, Ana-Marie Duta, Christopher Forbes, and Katherine Darroch, whom I have supervised on this topic in the project components of their respective degrees.

Introduction

The COPINE (Combating Paedophile Information Networks in Europe) classification model (Quayle 2008) rates the severity of victimization in child pornographic imagery on a ten-point scale (Taylor, Quayle and Holland 2001). In the United Kingdom (UK) and elsewhere in Europe, this scheme is used as an aid to investigation and criminal proceedings and has been adapted for use by the UK's Sentencing Advisory Panel as a five-point scale (Table 8.1) for use in court cases (*Regina* v *Oliver*, Court of Appeal, Criminal Division, 21 November 2002).

Such classifications afford a standard approach for gauging the seriousness of offenses, based upon the "strength" of the imagery possessed or exchanged by alleged offenders. In court, the scheme can assist in limiting the need to expose individuals – those attending or participating in the proceedings – to graphic, potentially distressing, or damaging materials. Instead, an outline description of the classification scheme, combined with the degree and quantity of the materials in a suspect's possession, can convey the significant dimensions in a criminal case, that is, the degree or "strength" of individual images or video content and the quantity of materials that fall under each classification.

Table 8.1   Sentencing Advisory Panel scale

1. Images depicting nudity or erotic posing with no sexual activity
2. Sexual activity between children, or solo masturbation by a child
3. Non-penetrative sexual activity between adult(s) and child(ren)
4. Penetrative sexual activity between child(ren) and adult(s)
5. Sadism or bestiality

For some professionals in law enforcement, legal representation, or digital forensics, exposure to hard-core pornography in the form of explicit images and video is often a necessary aspect of the job. Such individuals may be called upon to scrutinize or pass judgment on the materials with which they are confronted. Since they may be required to apply such classification schemes, they will not directly benefit from schemes of this type.

Another major area of concern for law enforcement in which content has a crucial role is the use of social networking and similar chat-based sites for the sexual grooming of minors. Through such networking sites, those wishing to engage in criminal exploitation of children may seek to develop a relationship with one or more minors as part of the grooming process. In such settings, the content may be entirely (or largely) text based and the collation of evidence against an offender will require the assembly and annotation of such computer-based interactions.

As in the cases of graphic imagery noted earlier, there are several contexts in which exposure to explicit text-based materials may be a professional requirement – as a digital forensic investigator, a legal professional, or a member of law enforcement. With this in mind, our purpose is to explore the possibility of developing techniques, based upon the classification of such textual content, which will provide means to neutralize by varying degrees the strength of content to which a student, trainee, or other professional may be exposed. Our aim is to manage the quantity and degree of such content with a view to minimizing any detrimental effects (observer impact) that such content may have on ill-prepared individuals.

Observer impact

Anyone who has encountered pornography is likely to be aware of its strong emotional force. Furthermore, this generally increases with the "strength" of the pornographic content. Although there is some dispute over the long-term psychological impact of such exposure (see, for example, Allen, D'Alessio and Brezgel 1995; Davies 1997; Malamuth, Addison and Koss 2000), there is strong evidence to suggest that such exposure may detrimentally affect the psychology of the individual. For example, Paolucci, Genuis, and Violato (1997) conclude that "the results are clear and consistent; exposure to pornographic material puts one at increased risk for developing sexually deviant tendencies, committing sexual offenses, experiencing difficulties in one's intimate relationships, and accepting the rape myth" (Paolucci et al. 1997, p. 3). Other sources echo this perspective for "strong" pornography and suggest that viewers of extreme pornographic images may develop tolerance to such images, and this may affect sexual response (see Segal 2014).

Individuals engaged in detection, evidence gathering, and legal proceedings involving child pornography face a risk of "observer impact," since work commitment means that distancing themselves from such materials becomes impossible. A 2010 study by Perez, Jones, Englert, and Sachau (2010) examined the "psychological impact of viewing disturbing media on investigators engaged in computer forensics work" and concluded that "substantial percentages of investigators reported poor psychological well-being. Greater exposure to disturbing media was related to higher levels of STSD [secondary traumatic stress disorder] and cynicism" (Perez et al. 2010, p. 113). The presence of such "side-effects" and the need to attend to their impact is clearly noted: "although law enforcement agencies need individuals to view potentially disturbing images for investigative purposes, there is little research examining the impact of such work on individuals' psychological well-being. This study was the first to assess quantitatively the levels of burnout and STSD among employees who are required to view disturbing media" (p. 120). In conclusion, this study offered "empirical data on the poor psychological well-being of law enforcement employees investigating computer child pornography cases in terms of both burnout and secondary traumatic stress" (p. 120).

Text-based content

We have seen the development of scales for the "strength" of pornographic images, and these scales have been used in research and for judicial purposes, where their application acknowledges the desire to avoid unnecessary exposure to such content. This focus makes sense, in light of the prevalence of "strong" imagery and its immediate impact upon an observer. In contrast, sexually explicit text content has not received the same attention. One reason for this disparity may be that textual content is considered less "forceful" than its graphical counterpart. Nevertheless, there are contexts in which exposure to explicit textual materials may be a professional requirement.

A major concern for law enforcement is the use of social networking and similar chat-based sites for the sexual grooming of minors. While such interactions may facilitate the creation or transfer of pornographic images and lead to child exploitation, the medium of interaction may be entirely (or largely) text. In order to pursue a criminal case against such grooming, law enforcement officers must collate evidence against a suspect. Inevitably, this will require the assembly and annotation of such textual content. Such material is often highly sexualized and explicit, so we have a context in which several individuals may be exposed to this content, with its associated "observer impact." Such a legal case may confront law enforcement officers, digital forensic investigators, legal professionals, and associated support staff with such materials. This should raise duty of care concerns toward these professionals for whom firsthand exposure to text-based sexually explicit or paedophilic materials is an obligation.

Gauging text content

In the research reported here, we have been exploring means to measure and neutralize sexually explicit textual content in the belief that through control of the quantity and degree of such content, there is scope to manage the detrimental effects of observer impact. With this objective in mind, a simple strategy may suggest itself. Since scales for graphical child exploitation materials already exist, could we simply apply COPINE or SAP as the basis for classifying the strength of textual content?

There is a key difference between the intended application of scales like COPINE and SAP on the one hand and scales for grading sexually explicit text on the other. Specifically, the former are directed toward ranking an offense and are brought to bear in cases where mere possession of the imagery may itself be illegal. Hence, application of the imagery scale facilitates judgment on the seriousness of a criminal offense. This has no precise parallel in the context of text content. There is no comparable offense associated with "possession" of sexually explicit text – to any degree of explicitness.

Still, the descriptions inherent in scales for graphical content could be applied directly to textual content. This would allow the possibility of grading texts according to the content that they describe. A textual description of child exploitation might thereby reflect a high (or low) level on the COPINE scale.

Despite the feasibility of this application, a serious deficiency remains with the notion of applying such a scale as the basis for grading the "strength" of sexually explicit text. The problem is that COPINE and SAP are solely focused on child exploitation, and while this area is most relevant for law enforcement, it excludes many aspects of sexual behavior and descriptions that have no bearing on underage individuals but may nonetheless cause concern, anxiety, or worse in those obliged to read such content. Evidently, the realm of application for COPINE and SAP does not extend to the full range of content that would reasonably give rise to the duty of care concern, as expressed above.

There are numerous techniques available for automatic text classification, and these vary in their approach and effectiveness (Weir and Duta 2012). Such approaches may be suitable for automatically recognizing that any specific text is sexually explicit. In seeking a means to gauge the degree of sexually explicit content, we have not addressed this identification issue but begin from the assumption that we know the text samples under consideration are sexually explicit to a degree that falls within the realm of concern.

In order to experiment with neutralization of explicit texts, we adopted a simple approach to gauging the degree of explicit text content by accumulating sample texts and creating a data set of sexually explicit terms. This was gathered through a survey of online materials, with initial data extracted from chat logs from the Perverted Justice website (www.perverted-justice.com). This was supplemented with data derived from Internet-sourced sex stories and online archives of text-based pornography. On this basis, we derived a list of sexually explicit vocabulary. Using this data set, we can match against sample texts in order to quantify the degree of sexually explicit content in the sample.

Impact reduction

Our approach to reducing the impact of sexually explicit text is based upon the idea of "obstructing" the immediate perception of the text content. One benefit of working in the text domain is that such content affords a wider degree of granular analysis and alteration than is possible with images.

For any sample text, we can employ a variety of matching tests to determine what specific lexical components contribute the explicit sexual content. At the simplest level, this requires string matching on individual words and multiword units. This can be supplemented with varieties of fuzzy matching, from word stemming, through concept clustering, to metaphor matching. In word stemming all variations on a root term are considered equal (e.g., fuck, fucking, fucker, fucked). Stemming simplifies the process of "weighing" the sexual content, for example, four instances of the same stem rather than four different terms. Concept clustering is another form of aggregation. In this case, instances of the same concepts are treated alike (e.g., fuck, screw, shag, etc.). Such clustering also simplifies the counting process. Thereby, we may register three instances of the same concept rather than three different terms. A further approach, proposed but yet to be implemented, employs metaphor matching. In this case, we may selectively match terms that serve as metaphors for sexual features. For example, some references to long, pointed instruments can be treated as penis metaphors. Embracing metaphor aims to capture innuendo as well as sexual connotations, such as "put my sausage in your oven."

Using such techniques, in combination with an annotated dictionary, we can isolate individual words, multiword units, and larger discourse components and selectively replace these aspects with alternative expressions. This is the basis for our process of neutralization.

Neutralization is how we term the procedure that gauges and applies a reduction in the emotive force of the sexual content. The aim of neutralization is to modify the content in a manner that can convey its "seriousness" and meaning without full exposure to the "raw" materials. In this regard, we are exploring the feasibility of such neutralization and strategies for effecting such change in sexually explicit texts. Based upon the estimation of sexually explicit content for any sample text, we aim to apply varying degrees of reduction in explicitness, usually maintaining the meaning but reducing the "impact."

To accommodate better the richness of possible sexual content, we include a further dimension in the estimation. We call this the "weight" of any sexually explicit lexical unit. This denotes the strength of word or expression. Thus, one sexually explicit term may be regarded as stronger than another (even if they have the same meaning). Distinguishing greater or lesser "weight" in source content allows for selective targeting of expressions with lesser or greater strength. Substitution based on levels of emotive force can be more sophisticated and flexible than a simple binary representation of explicit or not. Furthermore, this flexibility allows for greater variety in handling contexts and individuals for whom neutralization may be appropriate.

Implementation

Following the principles and approach described above, we have implemented a neutralization test facility that operates with three strength levels and four varieties of item replacement. The item replacement options are as follows:

1 Vowel/asterisk replacement
2 Euphemism
3 Full asterisk replacement
4 Floralization

Seeking to reduce the immediate impact of the explicit language, we have adopted one technique that is often used for obscuring words, the replacement of characters with asterisks. Two forms of this replacement are used (1 and 3 in our list). In case 1, the first vowel in the sexually explicit expressions is replaced with an asterisk. Displaying explicit texts with such replacements has a minimal effect but still slows down the reader's apprehension of the meaning and thereby reduces its impact. The second replacement strategy replaces sexually explicit terms or expressions with euphemistic alternatives. This is considered more effective in reducing the impact of the original text since there is a greater degree of substitution than with vowel/asterisk replacement. Once again, such replacement aims not to obscure the meaning of the original "raw" text.

A further increase in the level of substitution is achieved by applying full replacement of characters in sexually explicit expressions with asterisks. This has a greater obscuring effect but still tends not to prevent the reader from grasping the intended meaning of the original text (given that the reader appreciates the motivation behind the substitutions). The fourth replacement strategy adopts a radical approach to obscuring expressions in the original explicit texts. Floralization replaces target terms with the names of fragrant flowers to achieve a different form of cognitive distraction.

In our prototype system, this range of replacement strategies (neutralization options) is combined with our three strength levels to afford a "replacement matrix." We can select how any target item is replaced according to its strength. This permits us to "mix and match" replacement strategies to accommodate different individuals or different varieties of raw materials.

The system for exploring neutralization outputs XML marked-up text that indicates the strength of the source item and the type of neutralization for each change. Table 8.2 presents an example with alternative

Table 8.2   Example text with alternative replacement strategies

---

Original text
I want to rub your tits and your pussy until my cock explodes.

Asterisk 1
I want to rub your <L3.O2>t*ts</L3.O2> and your <L1.O2>p*ssy until my
<L2.O2>c*ck</L2.O2> explodes.

Euphemisms
I want to rub your <L3.O1>breasts</L3.O1> and your <L1.O1>vagina
</L1.O1> until my <L2.O1>penis</L2.O1> explodes.

Asterisk 2
I want to rub your <L3.O3>****</L3.O3> and your <L1.O3>*****</L1.O3>
until my <L2.O3>****</L2.O3> explodes.

Floralization
I want to rub your <L3.O4>blueberry</L3.O4> and your <L1.O4>poppy
</L1.O4> until my <L2.O4>campion</L2.O4> explodes.

Mixed (euphemism, floralization, asterisk 1)
I want to rub your <L3.O2>t*ts</L3.O2> and your <L1.O1>vagina</L1.O1>
until my <L2.O4>campion</L2.O4> explodes.

---

neutralization options and their combination. The XML tags indicate
the strength level of the targeted item and the replacement technique
applied.

Conclusions

In different contexts, there will be a need for different means of quanti-
fication. If we are operating with sets of documents, we may be satisfied
with a density ratio for each document or an average for documents
in the set. If we are tracking dynamic content, such as chat or social
media, we may need a dynamic measure of sexually explicit content
rather than, or in addition to, an aggregated density ratio. In other cases,
we may be especially interested in trends.

Documents may be gauged for the density of their sexually explicit
content, for example, by an aggregate measure, such as the ratio of sexual content, or by the proportion of sexual content to non-sexual content. A real-time context, such as ongoing interactive chat, may require
dynamic indicators, such as a sequence of measures for content that
facilitate generating a plot for progressive content. Whatever the basis
for quantifying the sexually explicit content in text, this may be used as
a driver for neutralization.

We have focused attention on the need for quantifying sexual content that is in textual rather than image form and have produced a test bed facility that serves as a means to explore the prospects, techniques, and benefits of neutralization.

Bibliography

Allen, M., D'Alessio, D. A. V. E. and Brezgel, K. (1995) "A Meta-Analysis Summarizing the Effects of Pornography II Aggression after Exposure," *Human Communication Research* 22(2): 258–83.

Davies, K. A. (1997) "Voluntary Exposure to Pornography and Men's Attitudes Toward Feminism and Rape," *Journal of Sex Research* 34(2): 131–7.

Malamuth, N. M., Addison, T. and Koss, M. (2000) "Pornography and Sexual Aggression: Are There Reliable Effects and Can We Understand Them?," *Annual Review of Sex Research* 11(1): 26–91.

Paolucci, E. O., Genuis, M. and Violato, C. (1997) "A Meta-analysis of the Published Research on the Effects of Pornography," *Medicine, Mind and Adolescence* 72(1–2).

Perez, L. M., Jones, J., Englert, D. R. and Sachau, D. (2010) "Secondary Traumatic Stress and Burnout Among Law Enforcement Investigators Exposed to Disturbing Media Images," *Journal of Police and Criminal Psychology* 25(2): 113–24.

Quayle, E. (2008) "The COPINE Project," *Irish Probation Journal*, 5 September 2008.

Regina v Oliver Court of Appeal Criminal Division, 21 November 2002, http://www.inquisition21.com/pca_1978/reference/oliver2002.html (accessed 10 October 2014).

Segal, D. (2014) "Does Porn Hurt Children?," *New York Times*, 28 March, http://www.nytimes.com/2014/03/29/sunday-review/does-porn-hurt-children.html (accessed 10 October 2014).

Taylor, M., Quayle, E. and Holland, G. (2001) "Child Pornography, the Internet and Offending," *The Canadian Journal of Policy Research* (ISUMA) 2(2): 94–100.

Weir, G. R. S. and Duta, A. (2012) "Strategies for Neutralising Sexually Explicit Language," *Proceedings of the Third IEEE. In Cybercrime and Trustworthy Computing Workshop (CTC)*, 66–74.

9

Spreading the Message Digitally: A Look into Extremist Organizations' Use of the Internet

Richard Frank, Martin Bouchard, Garth Davies, and Joseph Mei

Introduction

Why would a terrorist choose to utilize the Internet rather than the usual methods of assassination, hostage taking, and guerrilla warfare? Conway (2006) identified five major reasons why extremist groups used the Internet: virtual community building, information provision, recruitment, financing, and risk mitigation. Terrorist and extremist organizations can use the Internet to increase their visibility and provide information about the group along with its goals without posing an increased risk to the members. It also allows them to easily ask for, and accept, donations through anonymous financial services such as Dark Coins. These benefits allow these groups to promote awareness of their cause, to convey their message to, and perhaps foster sympathy from a much larger pool of potential supporters and converts (Weimann 2010). Finally, the Internet also provides asynchronous services with global access, with the sender and recipient located at any place, at any time, without the need to link up at a specific time (Wagner 2005). In short, unlike the real world, cyberspace is borderless without limitation, and this makes identification, verification, and attribution a challenge.

Today, almost all active terrorist organizations have recognized the benefits of the Internet, as they "maintain their own website, or have sites dedicated to them" (Saint-Claire 2011). Many of these contain more than one site under several different languages. Some, like ISIS, even utilize social media to further their propaganda. These groups come from all corners of the globe, including the Middle East (e.g., Hamas and Hezbollah), Europe (e.g., Basque ETA movement), Latin America (e.g., Armed

Revolutionary Forces of Columbia FRAC), and Asia (e.g., the Liberation Tigers of Tamil Eelam LTTE and Aum Shinrikyo) (Weimann 2006).

Many writers have discussed the role of the Internet for terrorist groups, but few empirical studies have analyzed the content of such websites (Bouchard, Joffres and Frank 2014; Chen 2012; Davies, Bouchard, Wu, Joffres and Frank, forthcoming; Seib and Janbek 2011; Weimann 2006 are some of the notable exceptions). One approach to analyze the content of violent extremist groups' websites is to manually visit the websites, read each webpage, find a way to code and summarize the content of each webpage, search for all links leading to other webpages, and finally map out the hyperlinks between those pages. Analyzing the website manually might lead to accurate conclusions about the content, but studying the structure and content of a large website with thousands of pages in this fashion is infeasible. Due to the large scale of the problem, the data collection and analysis must be performed by computers. This paper attempts to further this line of research through the presentation of an automatic data retrieval tool to capture data and analyze it for the sentiment expressed within. Results of studying the sentiment expressed on these websites are presented, and then the paper offers some final conclusions.

The Terrorism and Extremism Network Extractor (TENE)

Web-crawlers are the tools used by all search engines to automatically map and navigate the Internet and collect information about each website and webpage they visit. There are many off-the-shelf web-crawler solutions available on the Internet for purchase, such as Win Web Crawler (http://www.winwebcrawler.com), WebSPHINX (http://www.cs.cmu.edu/~rcm/websphinx), and Black Widow (http://sbl.net). Given a starting webpage, they will recursively follow the links from that webpage until some specified termination condition applies, capturing everything along the way. During this process the software will keep track of all the links between other websites and (optionally) follow and retrieve those as well. The content, as it is retrieved, is saved to the hard drive of the local user for later analysis. The purpose of most web-crawlers is to simply save the retrieved content onto the hard drive, to "rip" a webpage or website because it contains content desired by the end-user. More advanced analysis is usually not part of the package.

Terrorism and Extremism Network Extractor (TENE) is similar in spirit in that it also browses the World Wide Web, but since it is a custom-written computer program, it is more flexible. TENE specifically is capable of seeking out terrorism and extremism content based on user-defined keywords and other parameters. As it visits each page, TENE captures it for later analysis, while simultaneously collecting information about the content and making decisions on whether it is a page on the topic of extremism or not. The idea of this approach is based on a combination of the work associated with the Dark Web project (Chen 2012) at the University of Arizona and previous research to study online child exploitation websites at Simon Fraser University (Joffres, Bouchard, Frank and Westlake 2011; Frank, Westlake and Bouchard 2010; Westlake, Bouchard and Frank 2011).

Figure 9.1 shows an overview of TENE. TENE is actually a system that can be distributed across multiple machines, depending on the resources available. The first step (see Figure 9.1) is to define a job along with its parameters. TENE can handle multiple jobs simultaneously, each of which is given a priority. The priority of the job determines the relative resources allocated to it. For example, if jobs A, B, and C are given priority 50, 80, and 70, respectively, then job A will receive 25 percent $\{[50/(50+80+70)]*100\%\}$ of the available resources. If more computers are added to TENE, then the absolute amount of resources available

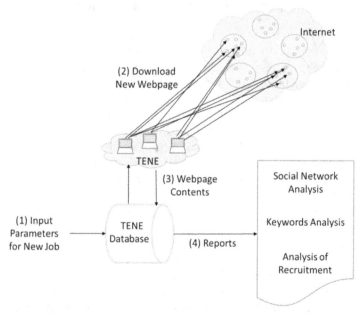

Figure 9.1   TENE overview

to each crawl is increased, although the relative amount of resources available for each job will remain the same.

Each "job" has four parameters imposed to prevent it from perpetually crawling the Internet and wandering into websites/webpages unrelated to extremists. The parameters are described in the sections below.

Number of webpages

For all practical purposes, the number of webpages on the Internet is infinite, so a limit on the number of webpages retrieved must be included to manage the extraction process time. Any crawler could theoretically crawl for a very long time and store the entire collection of webpages on the Internet; however, for this project this is infeasible for several reasons. First, the amount of storage required to store the results would be out of the scope of this research project. Second, webpages are created at a much higher rate than what can be retrieved with TENE. Finally, valid conclusions can be drawn from representative samples, and a copy of the "full Internet" is not required.

Number of domains

The number of domains across which TENE collects data must also be specified. When limiting a crawl to n pages, the crawler will attempt to distribute the sampling equally across all websites that it has encountered thus far, meaning that TENE will sample similar numbers of webpages from each of the websites. Thus, at the end of the task, if w websites were sampled, all websites will have approximately the same number of webpages retrieved ($=n/w$) and analyzed.

Trusted domains

There are a set of domains specified that tell the crawler that everything on that domain is trusted and that it should not retrieve any webpage on those domains. For example, www.microsoft.com can be assumed, with an extremely high likelihood, not to contain extremist material. Thus, without retrieving any pages from that website, TENE would assume that the website does not contain extremist content. Further, without having this mechanism available, TENE could wander into a search engine, leading it completely off topic and making the resulting network irrelevant to the specified topic.

Keywords

The purpose of TENE is to find, analyze, and map out the websites and webpages of extremists. To achieve this, TENE recursively retrieves all webpages that are linked from the webpage it is currently looking at. However, since the webpages that discuss extremism are a very tiny subset of all the webpages on the Internet, it would be expected that, unconstrained, TENE would very quickly start to retrieve webpages that are completely unrelated to extremists. Thus, some mechanism must be built into TENE that controls which webpages it uses in its exploration process. This is done through the use of keywords, which are user-specified words that have been found to be indicative of extremist webpages and thus indicate to TENE that the webpage currently being retrieved is on topic. Within the extremist domain, such keywords could include *gun*, *weapon*, or *terror*. To make this mechanism more robust, a counter can be specified which indicates a minimum threshold for TENE on the number of keywords that must exist on a page before it is considered on topic.

Once the job is created, each webpage is downloaded (Figure 9.1, Step 2). If it meets the parameters laid out above, then the page is considered "good," all content is saved and all links are followed out of it recursively (Step 3). The webpage contents are then part of the TENE database, and various reports and analysis can be performed (Step 4).

Sentiment

The use of keywords presents a good first step in trying to determine content (e.g., Bouchard et al. 2014; Davies et al. forthcoming). However, the use of single keywords may lead to misleading interpretations of content. For example, if *gun* and *control* are found within close proximity of each other, it might be concluded that the text is discussing *gun control*. This would most likely not indicate an extremist webpage but more likely that the webpage is written by a proponent or opponent of gun ownership and is thus not relevant in TENE's data collection. However, text containing *gun* and *control* could be discussing "controlling someone with a gun" within the domain of kidnapping, for example, in which instance TENE should continue the analysis. Thus, keywords can give an indication of the content within a webpage but cannot be used to determine exactly what that content is about. For a more accurate insight into the content of a specific piece of text, more powerful computational tools, such as sentiment analysis, are needed. This paper uses SentiStrength (Thelwall et al. 2010; Thelwall et al. 2013) (discussed below) to analyze the sentiment expressed on these websites. Long-term plans are for TENE to immediately and automatically determine whether each webpage it encounters is discussing extremism from a positive, negative, or neutral angle, and if relevant then continue the exploration using the hyperlinks found within the webpage.

Sentiment analysis

Sentiment analysis is a category of computing science that specializes in automating the process of data analysis by arranging data into discrete classes and sections (Feldman 2013; Kennedy 2012). Sentiment analysis has been used for customer review analysis for products (Feldman 2013) or social media platforms to detect attitudes toward events or products (Ghiassi, Skinner and Zimbra 2013). This process is very difficult since it's been estimated that 21 percent of the time even humans cannot agree amongst themselves about the sentiment within a given piece of text (Ogneva 2010), with some people not understanding subtle context or irony. Thus SA systems cannot be expected to have 100 percent accuracy (when compared to the opinions of humans), but as long as they provide insight into the author's motivations and sentiment (negative, positive, or neutral) toward certain events or actions, the incorporation of SA into TENE would allow for an increased accuracy of identifying extremist webpages.

Some researchers have compared different types of forums based on their levels of negativity (Chalothorn and Ellman 2012) whereas others have used sentiment analysis to examine extremist content in different languages (Abbasi and Chen 2005). This even extends to the social context surrounding extremist events; some researchers have developed methods that attempt to read and predict the social media response to these types of events (Burnap et al. 2014).

The work presented in this paper is similar in goal to the above, with TENE automatically needing to know the sentiment expressed within a certain webpage. To facilitate this, TENE downloads a given webpage, strips out the HTML markup from within it, and the information contained within is then available for sentiment analysis. Since the goal of the paper is not to push the boundaries of sentiment analysis algorithms, it was decided early on in the research that the off-the-shelf algorithm SentiStrength would be used.

SentiStrength

SentiStrength uses a so-called "lexical approach" which is based on human-coded data as well as various lexicons (Thelwall and Buckley 2013). Although the program is designed for short, informal online texts, there are features in place that allow longer texts to be analyzed (Thelwall and Buckley 2013). SentiStrength analyzes a text by attributing polarity values of either positive or negative to words within the text; these values are augmented by so-called "booster words" that can influence the positive or negative values assigned to the text as well as negating words, punctuation, and other features that are uniquely suited for studying an online context (Thelwall and Buckley 2013).

One of the features of SentiStrength is its ability to analyze the sentiment around any given keyword. For example, the phrase "I love apples but hate oranges" can be analyzed for the sentiment around *apples* (resulting in a positive outcome) as well as *oranges* (resulting in a negative outcome). The words around a given set of keywords are compared to the sentiment dictionary and their resulting values make up the total sentiment score for any given text.

To analyze a given piece of text for multiple keywords, multiple iterations must be done, with each iteration analyzing the same piece of text albeit with a different keyword (resulting in the sentiment toward that keyword). Due to the very specific nature of this command, it is necessary to input each form of a particular word being analyzed. For example, to analyze sentiment around the word *kill,* it is necessary to also analyze at the same time *kills, killing, killed,* and all other derivatives of the word *kill.* Since there are many keywords around which TENE needs to know the sentiment, thus for each webpage, multiple iterations of SentiStrength are applied, each one returning the detected sentiment around the specific keyword (and its derivatives). At the end, each text is linked to multiple sentiment values, one for each keyword. Those multiple values are then averaged to produce a single sentiment score for any given piece of text. A negative value implies negative overall sentiment, while a positive value implies overall support for the ideas expressed within.

Data collection

First a sample of websites were selected for retrieval by TENE, after which the parameters of the crawl were determined. This allowed the data to be collected for analysis.

Sample of websites under study

We selected a total of nine websites for analysis. In making selections, we were interested in finding websites that 1) were likely to host extremist content, 2) offered a diversity of themes and/or causes, and 3) had different purposes and/or target audiences (e.g., discussion forums, news sites, websites of militant groups). We drew from keyword searches and from a number of reports and prior studies on extremism (e.g., Bowman-Grieve 2009; Chen 2012; Karagiannis 2005; Jefferis 2011; Komissarov and Stepanov-Egiyants 2013; Nedoroscik 2010; Valentine 2010) to create a short list that would fit the above criteria. The list of websites analyzed in this chapter can be found in Table 9.1. In some instances, multiple websites were found to be leading to websites run by the organization, in which case all such websites were grouped as belonging to the organization and thus were included in the data collection. What constitutes "extremism" is inherently subjective, and we do not necessarily classify the selected websites as "extremist" websites or groups. For example, Islamic Awakening, one of the forums we analyze, contains a host of information of a general nature of interest to Muslims, and the majority of the content and exchanges would not qualify as extremism. Yet, some sections on the forums give rise to heated debates, with some interventions decidedly falling on the violent end of the spectrum. The extent to which website content qualifies as extremism is, in part, an empirical question that tools such as SentiStrength can help answer.

Parameters for the web-crawler

Since the focus was to retrieve a specific *set of websites*, the crawler was set to focus only on those websites and not wander around on the Internet looking for specific *content*. Thus, the parameters of the crawl were set in the ways described below.

Number of webpages

Since the number of webpages would be bounded and relatively small, no limit on the number of webpages was provided.

Number of domains

For the purposes of studying a specific set of domains, TENE was fixed to the domains shown in Table 9.1. This meant that the ability of TENE to extend the searches outside of the seed websites was disabled.

Keywords

Usually content is being sought out, and this is mainly driven by keyword searches. However, since this project is aiming to determine the sentiment expressed within a set of extremist websites, the goal was to capture everything and anything on those websites. Thus, keywords were not used to guide TENE during data collection.

Trusted domains

The trusted domains during the data collection performed for this project was not relevant since TENE was restricted only to those domains that were being collected.

Due to the nature of TENE, the sampling across each website was designed to be random, implying under ideal conditions the same number of webpages would be retrieved from each website. However, due to the size of the website, this is not always possible. For example, Army of God USA only contains a seeming four pages, although there is the possibility of it containing hidden or password-protected portions which TENE was not designed to seek out. Further, if a server (which hosts the website) is slow, data capture from it is going to progress more slowly than from a faster server, meaning more pages can be downloaded from a faster server, increasing the sample size from it. Given these two considerations, the number of webpages sampled from each server is also shown in Table 9.1. In total, 2,500 webpages were retrieved.

After stripping the HTML from each of the webpages, and running them through SentiStrength, the sentiment around each keyword was calculated. The average sentiment expressed on each website is shown in Table 9.2, along with the number of pages sampled from each website.

As can be seen in Table 9.2, some websites have no sentiment on certain keywords. For example, StormFront has no sentiment when looking at the keyword *terrorism*, which is to be expected, as the website is about white supremacy, not causing terror. This is in line with the negative sentiment regarding *war* (and its derivatives), which, in a slightly different sense, is even expressed in their slogan "We are the voice of the new, embattled White minority!" This sentiment is seemingly spread throughout the forum, which was then detected by SentiStrength. Interestingly, the sentiment toward Islam or Pakistan was nonexistent, while the sentiment toward America was, surprisingly, slightly negative.

The English sub-site of the Hizb ut-Tahrir website, an organization against American military action, seemed to have very strong, albeit all negative, sentiment toward *war*, *terrorism*, *attacks*, and surprisingly even *Islam*, but the most negative sentiment expressed on the website was actually toward *America*, or *Americans*. This could be interpreted as the group being very negative toward America, while at the same time also not condoning war, attacks, or terrorism. Though the Arabic sub-site of the Hizb ut-Tahrir website was also included in the study, it seems that SentiStrength was not at all able to deal with the lack of English words on the site, confirming one of the weaknesses of this program's ability to deal with non-English content.

The list of keywords was chosen based on the likelihood that they would appear somewhere in the text. Some were expected to be found rarely because of the varied nature of the domains; for example, keywords such as *ISIS* and *jihad* were detected less frequently than others. However,

Table 9.1 Websites retrieved by TENE

Name	Website	Description	Pages Sampled
Stormfront	www.Stormfront.org	A Utah-based white supremacist forum that focuses on white nationalism.	201
Hizb ut-Tahrir	english.hizbuttahrir.org/ arabic.hizbuttahrir.org forum.hizbuttahrir.org	An Islamic political organization that seeks the dissolution of the State of Israel and the re-establishment of the Caliphate.	1066
Army of God USA	www.aogusa.org	An anti-abortion Christian extremist group.	4
Islamic Awakening	www.islamicawakening.com forums.islamicawakening.com	An Islamic forum which hosts a variety of themes and exchanges, including radical views from some users.	554
The American Nazi Party	www.americannaziparty.com	A political extremist group which believes the United States is deteriorating because of racial diversity.	15
Life Site News	www.lifesitenews.com	A Christian news site that contains articles with radical views surrounding legalized abortion.	200
Shahmat	shahamat-english.com shahamat-farsi.com	A news site that publishes articles supporting violent jihadists.	201
Kavkaz Center	www.kavkazcenter.com/eng/ www.kavkazcenter.infowww.kavkazcenter.net	A news website for the Kavkaz Center, an organization that supports the mujahedeen and seeks to establish an Islamic state in Russia.	200
The Muslim Brotherhood	www.ikhwanweb.com	A group some countries have labeled as terrorists. They claim to renounce violence and seek democratic elections.	59

Table 9.2 Average sentiment score for each website

Domain	# of web pages	Keyword(s) around which sentiment is built									
		Islam, Islamic	ISIS	jihad, jihadist	government, governments, governmental	war, wars, warring	terrorism	attack, attacked, attacking, attacks	target, targets, targeting, targeted	Pakistan	American, Americans, America
english.hizbuttahrir.org	87	-2.736	-0.736	-4.529	-3.046	-7.770	-2.667	-4.080	-0.230	-0.322	-11.931
forum.hizbuttahrir.org	865	-11.177	-0.027	-1.431	-2.489	-10.342	-2.761	-9.286	-0.669	-2.383	-3.829
forums.islamicawakening.com	201	-19.060	-0.313	-4.746	-3.060	-14.900	-5.677	-21.358	-1.562	-11.811	-8.308
shahamat-english.com	75	-4.187	-0.040	-1.000	-1.120	-17.027	-0.173	-86.987	-0.413	0.000	-4.400
shahamat-farsi.com	126	0.000	0.000	0.000	0.000	0.000	0.000	0.000	0.000	0.000	0.000
www.americannaziparty.com	15	0.000	0.000	0.000	-3.400	-7.667	-1.067	-1.200	-0.133	0.000	-7.933
www.aogusa.org	4	0.000	0.000	0.000	-44.250	-27.750	-1.500	0.000	-2.000	0.000	-34.500
www.ikhwanweb.com	59	-0.814	-0.780	0.000	-3.254	-1.424	-9.051	-3.017	-0.475	-0.068	-0.407

www.islamicawakening.com	353	-2.326	0.000	-0.657	-1.000	-4.581	-1.405	-2.595	-0.150	-0.606	-0.822
www.kavkazcenter.com	117	-8.325	0.000	-0.462	-0.889	-5.821	-1.393	-1.231	-0.034	0.000	-3.376
www.kavkazcenter.info	73	0.000	0.000	0.000	0.000	-0.041	0.000	0.000	0.000	0.000	-0.068
www.kavkazcenter.net	10	0.000	0.000	0.000	0.000	0.000	0.000	0.000	0.000	0.000	0.000
www.lifesitenews.com	200	-0.035	-0.295	0.000	-8.380	-1.900	0.000	-2.845	-0.280	0.000	-0.640
www.stormfront.org	201	0.000	0.000	0.000	-0.194	-8.731	0.000	-0.279	-0.060	0.000	-0.711
Grand Total	2500	-6.360	-0.103	-1.179	-2.284	-7.456	-1.996	-8.434	-0.443	-1.872	-3.037

the intention behind choosing some of these more narrow words was that when they were found they would likely have very polarized sentiment. This was true for *jihad* as it registered one of the most negative sentiment scores (-276, in a thread on *forum.hizbuttahrir.org*) but the *ISIS* keyword's lowest value was –35 (on a page of *english.hizbuttahrir.org*) despite the polarizing opinions surrounding it. Conversely, the keywords which more consistently showed sentiment were broad terms like *government* and *war*; *government* likely because it encompasses individuals talking about a diverse set of governments from different countries and *war* due to the common practice of extremist groups claiming they are either at war with a particular other group or said group has started a war against them.

Figure 9.2 shows the distribution of sentiment scores across the entire set of webpages sampled. While there were some pages with positive sentiment values within this decidedly negative sample of websites, there were no major positive outliers. For example, the lowest sentiment value found was *war* with a sentiment score of –410, in comparison to the highest positive value which was *Islam* and its derivatives with 26. Overall, of the 25,000 sentiment scores, only 481 were positive, with 19,306 being 0, and 2,817 between –1 and –9, with the rest (2,396) being lower than –9. This shows clearly that these sites match up with their reputations for being more in line with negative attitudes and discussion. In addition, it also seems that the Islamic forums have the lowest scores on average. This is likely because they are much less filtered, in that the users who post are not acting as the professional face of an

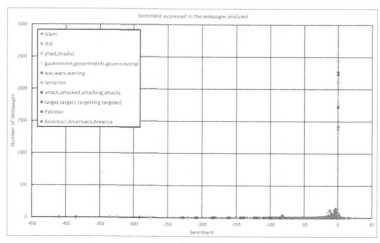

Figure 9.2   Distribution of sentiment scores across the pages sampled

organization like many of the news sources. This means they can utilize more extreme language and voice more extreme opinions.

Unfortunately, there were several pages which registered no sentiment score. There are a few possible reasons for this. First, some pages simply did not interact well with the sentiment software. For example, some of the Farsi pages from Shahamat were collected and analyzed, but the sentiment software is focused on examining exclusively English language documents and cannot assign values to different languages, especially ones with different character sets. Additionally, some of the links that the crawler followed linked to non-readable materials like .pdf documents that cannot be used by the crawler and thus cannot be interpreted by the sentiment software. Finally, there were some issues with simply accessing the pages themselves. There was a small portion of pages that were unable to be collected because of technical issues like incorrect SQL queries on the websites themselves. While not a significant portion, these lost results expose some of the weaknesses in the approach.

Discussions and conclusions

Previous studies have shown that with greater proliferation of the Internet and computing technology, the rates of individuals posting extremist content increases in turn (Holt 2012). To keep abreast of the growth of this online content, we must find alternative ways to study them. This deluge of data requires a method which can disregard the superfluous data that is often collected throughout an investigation. As these problems compound, automation has become an even more attractive alternative.

This study sought to examine a method to classify and interpret extremist content by combining sentiment analysis with the TENE webcrawler, with the long-term goal of integrating the sentiment analysis directly into the crawling process. For this paper, though, 2,500 webpages from 9 websites were retrieved and analyzed for the sentiment expressed within their content. It was found that the vast majority of the webpages have a sentiment score of 0, but if there is a sentiment score, it is very likely to be negative. This is in line with expectations.

This type of sentiment analysis would be very interesting when applied to the forum postings of individuals over time. It could aid in studying lone wolf extremists on web forums as well as the general tendencies of particular communities toward violent extremism. To do this would require similar steps to the current study. It would require the identification of terms associated with extremist attitudes and the development of the sentiment surrounding these terms over time. This could shed a light onto how violent extremism develops among these communities and individuals.

The study presented in this paper used keywords which were manually derived after a study of the webpages from which data was collected. These keywords were then used as part of the input in the SA algorithm in order to determine sentiment around those keywords. However, the keywords themselves could have been picked through a biased lens, possibly affecting the sentiment scores that were derived. In order to be exhaustive and impartial, Natural Language Processing tools could be used to identify all the nouns (and/or verbs and other types of language structures), and then those words fed into the SA. The use of NLP was not possible at the time of this study, hence this is left for future work.

A crucial development for web-crawlers such as TENE is to differentiate between websites supporting violent extremism and those against. Sentiment scores alone are unable to distinguish between websites arguing strongly for or against a similar cause. To determine whether SA can be used for the automatic identification of extremist websites, anti-extremist websites would need to be sampled, the sentiment extracted, and then some form of classification algorithm run to determine differentiating rules between the two types of webpages.

Bibliography

Abbasi, A. and Chen, H. (2005) "Applying Authorship Analysis to Extremist-group Web Forum Messages," *Intelligent Systems* 20(5): 67–75.

Apache Software Foundation (2014) "openNLP," http://opennlp.apache.org/ (accessed 30 December 2014).

Bouchard, M., Joffres, K. and Frank, R. (2014) "Preliminary Analytical Considerations in Designing a Terrorism and Extremism Online Network Extractor," in *Computational Models of Complex Systems*, V. Mago and V. Dabbaghian (eds.), New York: Springer, 171–84.

Bowman-Grieve, L. (2009) "Anti-abortion extremism online." *First Monday*, October. Available at http://journals.uic.edu/ojs/index.php/fm/article/view/2679.

Burnap, P., Williams, M., Sloan, L., Rana, O., Housley, W., Edwards, A., Knight, V., Procter, R. and Voss, A. (2014) "Tweeting the Terror: Modelling the Social Media Reaction to the Woolwich Terrorist Attack," *Social Network Analysis and Mining* 4(206): 1–14.

Chalothorn, T. and Ellman, J. (2012) "Using SentiWordNet and Sentiment Analysis for Detecting Radical Content on Web Forums," in the 6th Conference on Software, Knowledge, Information Management and Applications (SKIMA 2012), 9–11 September, Chengdu University.

Chen, W. (2012) *Dark Web: Exploring and Data Mining the Dark Side of the Web*, New York: Springer.

Conway, M. (2006) "Terrorism and the Internet: New Media, New Threat?," *Parliamentary Affairs* 59.

Davies, G., Bouchard, M., Wu, E., Joffres, K. and Frank, R. (forthcoming 2015) "Terrorist Organizations' Use of the Internet for Recruitment," in *Social Network,*

Terrorism and Counter-terrorism: Radical and Connected, M. Bouchard (ed.), New York: Routledge.

Feldman, R. (2013) "Techniques and Applications for Sentiment Analysis," *Communications of the ACM* 56(4): 82–8.

Frank, R., Westlake, B. and Bouchard, M. (2010) "The Structure and Content of Online Child Exploitation Networks," Proceedings of the Tenth ACM SIGKDD Workshop on Intelligence and Security Informatics 2004.

Ghiassi, M., Skinner, J. and Zimbra, D. (2013) "Twitter Brand Sentiment Analysis: A Hybrid System Using n-gram Analysis and Dynamic Artificial Neural Network," *Expert Systems with Applications* 40: 6266–82.

Holt, T. (2012) "Exploring the Intersection of Technology Crime, and Terror," *Terrorism and Political Violence* 24: 337–54.

Jefferis, J. (2011) *Armed for Life: The Army of God and Anti-Abortion Terror in the United States,* Santa Barbara, CA: Praeger.

Joffres, K., Bouchard, M., Frank, R. and Westlake, B. (2011) "Strategies to Disrupt Online Child Pornography Networks," *Proceedings of the EISIC - European Intelligence and Security Informatics,* Athens, September, 163–70.

Kennedy, H. (2012) "Perspectives on Sentiment Analysis," *Journal of Broadcasting and Electronic Media* 56(4): 435–50.

Komissarov, V. S., and Stepanov-Egiyants, V. G. (2013) "The Anti-terrorism Legislation of the Russian Federation." *Megatrend Review* 10(1): 335–49.

Ogneva, M. (2010) "How Companies Can Use Sentiment Analysis to Improve Their Business," http://mashable.com/2010/04/19/sentiment-analysis/ (accessed 28 September 2013).

Pang, B., Lee, L. and Vaithyanathan, S. (2002) "Thumbs Up? Sentiment Classification Using Machine Learning Techniques," *Proceedings of the Conference on Empirical Methods in Natural Language Processing (EMNLP),* 79–86.

Saint-Claire, S. (2014) "Overview and Analysis on Cyber Terrorism," http://www.iiuedu.eu/press/journals/sds/SDS_2011/DET_Article2.pdf (accessed 30 December 2014).

Sebastiani, F. (2002) "Machine Learning in Automated Text Categorization," *ACM Computing Surveys* 34(1): 1–47.

Seib, P. M. and Janbek, D. M. (2011) *Global Terrorism and the New Media: The Post-Al Qaeda Generation,* London: Routledge.

Thelwall, M., Buckley, K., Paltoglou, G., Cai, D. and Kappas, A. (2010) "Sentiment Strength Detection in Short Informal Text," *Journal of the American Society for Information Science and Technology* 61(12): 2544–58.

Thelwall, M. and Buckley, K. (2013) "Topic-based Sentiment Analysis for the Social Web: The Role of Mood and Issue-related Words," *Journal of the American Society for Information Science and Technology* 64(8): 1608–17.

Tong, S. and Koller, D. (2002) "Support Vector Machine Active Learning with Applications to Text Classification," *Journal of Machine Learning Research* 2: 45–66.

Wagner, A. (2005) "Terrorism and the Internet: Use and Abuse," in *Fighting Terror in Cyberspace,* M. Last and A. Kandel (eds.), http://www.worldscibooks.com/etextbook/5934/5934_chap1.pdf (accessed 30 December 2014).

Weimann, G. (2006) *Terror on the Internet: The New Area, the New Challenges,* Washington, DC: United States Institute of Peace Press.

Weimann, G. (2010) "Terror on Facebook, Twitter, and Youtube," *Brown Journal of World Affairs* 16(2): 45–54.

Westlake, B., Bouchard, M. and Frank, R. (2011) "Finding the Key Players in Child Exploitation Networks," *Policy and Internet* 3, Article 6: 1–32.

10

Criminals in the Cloud: Crime, Security Threats, and Prevention Measures

Alice Hutchings, Russell G. Smith, and Lachlan James

Acknowledgments

This chapter is a revised version of the authors' earlier work (Hutchings, Smith and James 2013), "Cloud Computing for Small Business: Criminal and Security Threats and Prevention Measures," originally published in *Trends and Issues in Crime and Criminal Justice* (Canberra: Australian Institute of Criminology). It is included in the current collection with permission. The research was supported by the former Department of Broadband, Communication and the Digital Economy and undertaken at the ARC Centre of Excellence in Policing and Security and the Australian Institute of Criminology. The opinions, findings, and conclusions or recommendations expressed are those of the authors and do not reflect those of the aforementioned agencies.

Introduction

Cloud computing refers to the delivery of computer processing infrastructure, operating systems, software, and data storage over Internet-based public or private computer networks (Mell and Grance 2011). Cloud computing can relieve users of some of the burdens associated with maintaining computers and data storage while enabling the associated costs to be reduced (Mowbray 2009). The range of cloud computing services (including some that are provided free of charge) that meet the particular needs of consumers is vast and growing rapidly.

This paper presents the findings of a systematic, desk-based assessment of English-language, public source literature available over the

Internet and through subscription-based services to identify current and emerging cloud computing risks and incidents. These have been categorized as crime and security risks that could affect cloud service providers, cloud computing tenants, and the transmission of data between cloud providers and their tenants. Many of these vulnerabilities are not unique to cloud computing but could arise in connection with conventional use of computer systems. What is different, however, is the nature of the data that may be stored in the cloud and its attractiveness to offenders located in disparate countries. Other vulnerabilities are unique to virtualization and the multi-tenancy environment of the public cloud.

Crime and security risks in the cloud

Compared with computer security incidents affecting corporate systems generally, there have been relatively few attacks reported against cloud service providers. Banham (2012) claims that this is because cloud service providers have stronger security, as they are more concerned about the consequences of reputational damage if data are breached. However, according to Pacella (2011, p. 70), "cloud providers often ask their clients to keep attacks quiet." When not bound by mandatory data breach reporting requirements, as in Australia, cloud computing tenants are likely to want to avoid the publicity associated with data breaches. This may be particularly relevant where the cloud service provider claims no responsibility or liability for breaches of data security or unavailability of data in service-level agreements (Blumenthal 2011). In addition, some attacks may go undetected, and in other cases when data breaches are made public, the fact that data were held in the cloud may not be released. It is clear that cloud service providers and vendors are reluctant to publicize the insecurity of their systems and are unwilling to disclose security breaches that occur. However, recent survey research involving cloud tenants, cloud service providers, and those who attempt to breach systems has revealed concerning evidence and actual victimization in connection with cloud computing.

Aleem and Sprott (2013) interviewed 200 ICT professionals worldwide. Respondents' most cited concern regarding the use of cloud computing was security, as reported by 93.4 percent of interviewees. This was considered to be more of a concern than governance (62.3%) and the lack of control over service availability (55.7%). Cloud computing had been implemented by 17 percent of respondents' organizations, and 8 percent of respondents reported that they had experienced a security breach in the cloud. Respondents reported that the top two cloud threats were data loss and leakage (73.5%), and account, service, and traffic hijacking (60.8%).

According to an international survey of 1,200 companies with over 500 employees in the United States, Canada, the United Kingdom, Germany, Japan, and India conducted by Trend Micro (2011), an anti-virus and computer security vendor, 43 percent of organizations that reported using cloud computing had experienced a "security lapse or issue" in the previous 12 months, although the nature of the incident was not disclosed.

Another survey of 103 United States and 24 European cloud service providers by the Ponemon Institute (2011) revealed that 62 percent were not confident that the cloud applications and resources they supply are secure. Sixty-five percent of respondents were public cloud providers, while private and hybrid cloud providers comprised 18 percent of the sample each. Of the public cloud providers, only 29 percent were confident or very confident that the cloud applications and resources supplied by their organization were secure.

A survey of 100 ICT professionals conducted by a computer security vendor at DefCon (a hacker conference held in Las Vegas in 2010) found that 96 percent of respondents believed that the "cloud would open up more hacking opportunities for them," and 45 percent had "already tried to exploit vulnerabilities in the cloud" (Fortify 2010, p. 7).

Crime and security risks involving cloud service providers

Authentication issues

Unauthorized access to cloud computing systems may occur when a username and password combination has been obtained without authorization. This can occur using a variety of technical and non-technical methods. Social engineering may be targeted toward the cloud service provider by, for example, claiming that urgent access is required but that the password is not working and needs to be reset. Passwords may also be guessed, be written down and left lying around, obtained using key-logging malware, cracked using brute force, or overcome when there are weak password recovery mechanisms, such as answering "secret" questions where the answers are publicly available (Dlodlo 2011). An example of a social engineering attack is provided in Box 10.1.

Box 10.1 Social engineering attack

zzzreyes writes:

I got an email from my cloud server to reset the admin password, first dismissed it as phishing, but a few emails later I found one from an admin telling me that they had given a person full access to my server and revoked it, but not before 2 domains were moved from my account. I logged into my account to review the activity and found the form the perpetrator had submitted for appointment of new primary contact and it infuriated me, given the grave omissions. I wrote a letter to the company hoping for them to rectify the harm and they offered me half month of hosting, in a sign of good faith. For weeks I've been struggling with this and figure that the best thing to do is to ask my community for advice and help, so my dear slashdotters please share with me if you have any experience with this or know of anyone that has gone through this. What can I do?

Source: http://it.slashdot.org/story/12/04/04/1738220/ask-slashdot-my -host-gave-a-strangeraccess-to-my-cloud-server-what-can-i-do?

Inadequate authentication checks may not necessarily be attributed to malicious activity, although they may result in data being accessed for nefarious purposes. For example, some incidents in which others are permitted to have access to data stored on the cloud may be inadvertent:

> In June 2011, Dropbox, a popular cloud storage site where approximately 25 million people store their videos, photos, documents, and other files, inadvertently left the site open for four hours on Father's Day. The glitch let anyone log in to customers' accounts with any password (Wright 2011, p. 20).

Insufficient or faulty authentication checks may also provide opportunities for Uniform Resource Locator (URL) guessing attacks, whereby possible page links are entered to gain access to pages directly, bypassing authentication checks (Grobauer et al. 2011). Data breaches, however they are accomplished, may have significant impact not only on the cloud service provider's tenant, whose data have been accessed, but also on customers who may have trusted that organization with their personal information:

> In early April, Epsilon announced that its database had been breached by an unknown third party, allowing unauthorised access to the email addresses of its clients' customers (Kunick 2011, p. 18).

Denial of service attacks

Denial of service (DoS) attacks against cloud service providers may leave tenants without access to their accounts. This can occur by sending a flood of traffic to overwhelm websites to make them inaccessible to legitimate users. When a DoS attack is conducted using a botnet (a network of compromised machines), this is referred to as a distributed denial of service attack, or DDoS. DoS attacks aimed at individual accounts, rather than at all cloud tenants, may also be accomplished by changing the tenant's password or maliciously continuing to enter the incorrect password so that the account becomes locked.

Use of cloud computing for criminal activities

Cloud computing accounts can be created or existing accounts compromised for criminal purposes. New cloud computing accounts may be created with stolen credentials and credit card details, thereby reducing the cost to the offender(s), as well as anonymizing the offender and creating further difficulties in tracking down the source of the attack, particularly when jurisdictions are crossed. Accounts created or compromised in such a way can be controlled as part of a botnet. In the following example, an existing cloud computing account was compromised and used to run a botnet command and control server:

> Computer World UK reported in December 2009 that a website hosted on Amazon EC2 had been hacked to run the Zeus botnet's command-and-control infrastructure (Blanton and Schiller 2010, n.p.).

Botnet command-and-control servers can be used to launch DDoS attacks, conduct scams such as click fraud, and distribute spam. The processing power of botnets may also be used to conduct brute force attacks to overcome password restrictions. For example, there have been reports that hackers made use of a cloud computing server to launch attacks on Sony's payment platforms in April 2011. This attack resulted in the breach of the personal data (including name, date of birth, and email address) of 77 million users across the globe, and it was believed that the data of around 11 million credit cards may also have been leaked (Hong Kong Government News 2011).

Cloud computing services may be used for the storage, distribution, and mining of criminal data such as stolen personal information or child exploitation material (Cloud Security Alliance 2010). Accounting systems run in the cloud may be attractive for money laundering and terrorism financing activities. The use of cloud computing to conduct illegal activities has had further negative consequences in relation to data access for other legitimate users of the cloud service provider when servers have been seized by a law enforcement agency. Not only may access be disrupted, but the law enforcement agency (international or domestic) may have access to that data in a multi-tenanted environment (Allen 2010).

Illegal activity by cloud service providers

Loss of access to data has also occurred when cloud service providers themselves have allegedly engaged in illegal behavior. In one case, the cloud service provider's services were stopped due to police action. Kim Dotcom and his colleagues were arrested following allegations the Megaupload cloud storage service he conducted involved illegal piracy. While the service had allegedly been used to swap movies and music files, Megaupload was also a low-cost way of legitimately sharing files and making online backups. When the authorities in the United States closed the service without warning, businesses were unable to access their documents (Bennett 2012).

Attacks on physical security

Cloud service providers' data centers may also be physically attacked, resulting in hardware theft, unauthorized access to servers, or loss of access to data. In one case, two masked men allegedly pistol-whipped a lone staff member during a graveyard shift, holding the worker hostage for two hours while confiscating equipment in a Chicago data center. The burglars reportedly entered the facility through a fire escape, swiped the staffer's access card through a reader, and forced him to perform a fingerprint scan before stealing computer storage equipment (Knapp et al. 2011).

Insider abuse of access

Cloud service provider insiders, such as employees, contractors, or third-party suppliers, may misuse their privileges to disrupt access or to obtain unauthorized access to stored data. Insiders may obtain employment at a targeted cloud service provider, be targeted by organized crime syndicates, abuse their access as the result of becoming discontent in their employment, or become tempted by presented opportunities and the potential perceived gains (Blumenthal 2011). Over half (52.9%) of the ICT professionals surveyed by Aleem and Sprott (2013) indicated that they were concerned about insider threats in the cloud.

Malware

The cloud service providers' servers may be vulnerable to malware infection, including virtual machine-based rootkits (Aron 2011). These risks apply in non-cloud environments as well. Malware infection may result in account names and passwords being compromised, files being accessed and copied, corruption of files, or being added to a botnet. There is also the possibility that malware compromising one tenant's virtual machines could then spread to the virtual machines of other tenants.

Side channel attacks or cross-guest virtual machine breaches

"Side channel attacks" or "cross-guest virtual machine breaches" may result in tenants crossing the shared virtual machine boundaries and accessing the data of other tenants using shared physical resources. Side channel attacks require the attacker's virtual machine and victim's virtual machine to be located on the same physical machine; therefore, these attacks may be random and not targeted toward a specific tenant. However, targeted attacks may still be possible. It has been demonstrated with one cloud computing service provider that co-tenancy could be successfully achieved 40 percent of the time by setting up new accounts while simultaneously manipulating the resource needs of the targeted victim's virtual machines (Ristenpart et al. 2009). Vulnerabilities in shared technology resources, which is the environment where side-channel attacks occur, was listed as one of the top cloud threats by 37.3 percent of the ICT professionals surveyed by Aleem and Sprott (2013).

Vulnerabilities in software applications

Security holes and vulnerabilities may exist in software applications run in the cloud, such as backdoors that bypass normal authentication protocols (Pacella 2011). New vulnerabilities for operating systems, Internet browsers, and business applications are regularly being identified. Security patches fix vulnerabilities in computer programs, which may be used to gain unauthorized access. However, delays in installing patches may lead to increased exploitation attempts as the vulnerabilities they are fixing are then made known.

Similarly, insecure application programming interfaces, which allow software applications to interoperate with each other by passing login information between them, may provide another attack vector. Of the ICT professionals surveyed by Aleem and Sprott (2013), 39.2 percent indicated that insecure application programming interfaces were among the top cloud threats.

Web browsers, used to access the Internet and cloud service providers, are another type of software application that may be subject to attack. Browser vulnerabilities include cross-site scripting whereby code is injected into websites and executed by the browser (Dlodlo 2011). Cross-site scripting can be used to hijack sessions by obtaining cookies or authentication credentials by redirecting users to a site impersonating the cloud service provider.

Cryptanalysis of insecure or obsolete encryption

Data stored in the cloud may be encrypted to prevent it from being read if accessed without authorization. However, encryption can potentially be weakened or broken if insecure or obsolete (Grobauer et al. 2011). Partial information can also be obtained from encrypted data by monitoring clients' query access patterns and analyzing accessed positions (Agrawal et al. 2011).

Structured Query Language injection

Structured Query Language (SQL) is a programming language used for database management systems. SQL injection attacks targeting web entry forms involve inputting SQL code that is erroneously executed in the database back end (Dlodlo 2011). SQL injection attacks can result in data being accessed and modified without authorization. Another injection attack is OS injection or command injection, whereby the input contains commands that are erroneously executed by the operating system (Grobauer et al. 2011).

Crime and security risks targeting cloud computing tenants

Phishing

Although, in the context of cloud computing, phishing misrepresents the provider, the attack is directed toward those who may hold an account with that organization with the aim of obtaining passwords and other identifying information to obtain unauthorized access to data held in the cloud (Dlodlo 2011). Phishing is one example of social engineering in which an email appearing to be from a legitimate organization is sent directing recipients to a bogus (spoofed) website to enter their login credentials or other personal information.

Domain name system attacks

Cloud computing users may be subject to domain name system (DNS) attacks. The principal use of domain names is to convert an Internet protocol resource (a string of numbers) into a readily identifiable and memorable address, such as those used in email addresses and URLs. A variety of DNS attacks are aimed at obtaining authentication credentials from Internet users, including cloud service tenants. Pharming and DNS-poisoning involve diverting visitors to spoofed websites by "poisoning" the DNS server or the DNS cache on the user's computer. Domain hijacking refers to stealing a cloud service provider's domain name, while domain sniping involves registering an elapsed domain name. Cybersquatting refers to registering a domain name that appears to be similar to a cloud service provider, which can be used to conduct phishing scams. Login details can also be obtained by typesquatting, which relies on a user entering the wrong URL and subsequently providing their authentication credentials to a spoofed website.

Compromising the device accessing the cloud

Access to a cloud computing account may be achieved if the device accessing the cloud service is compromised, for example, by a key-logger that records keystrokes including usernames and passwords (Banks 2010). Again, malware infections such as this may be random, or individuals may be directly targeted with a Trojan – malware designed to look like a legitimate file.

Access management issues

Businesses that fail to restrict their employees' access to cloud computing services after they leave their employment would be vulnerable to having their data accessed, altered, copied, or deleted (Subashini and Kavitha 2011). Former employees may seek revenge against their employer, or steal information for resale or to use in setting up a competing business. Such risks, of course, also exist in the non-cloud computing environment.

Attacks targeting the transmission of data

Session hijacking and session riding

Session hijacking involves the attacker exploiting active computer sessions by obtaining the cookies that are used to authenticate users. This can be achieved by cross-site scripting, which involves malicious code being injected into the website, which is subsequently executed by the browser (Dlodlo 2011).

A similar attack called "session riding" is where websites are exploited using cross-site request forgery to transmit unauthorized commands. An attacker "rides" an active computer session by tricking a user (for example, by sending a link) into visiting a manipulated webpage while they are logged into the targeted site. The webpage contains a request that is executed by the website as the user is also sending their authentication credentials. Commands may be used to, for example, manipulate or delete data, reset passwords, add new users or delete existing users, or forward emails (Schreiber 2004).

Man-in-the-middle attacks

In a man-in-the-middle attack, the attacker intercepts traffic between a website and a browser (Grobauer et al. 2011). This occurs when the browser believes that the attacker is the legitimate website and the website authenticates the attacker as the browser. The attacker can then read and alter the data being transmitted, including account passwords that may be used to log in to cloud services.

Network/packet sniffing

Network or packet sniffing involves the interception and monitoring of network traffic (Subashini and Kavitha 2011). Data that are being transmitted across a network, such as passwords, can therefore be captured and read if not adequately encrypted. In the cloud environment, this is particularly important as passwords play a critical role in establishing access to the provider's services.

Responding to crime and security threats in the cloud

Although the threats identified above may seem somewhat oppressive, there are a number of measures that can be adopted by cloud users, as well as cloud providers, to detect, prevent, and minimize the damage from criminal and security threats in the cloud environment. Table 10.1 provides an overview of the various methods that are available both to business tenants and cloud computing providers.

Table 10.1 Summary of cloud computing crime prevention and security measures

Crime and security risks involving cloud service providers	Technical prevention measures						Physical security					Organizational policies, awareness, and training					
	Patching operating systems, Internet browsers, and software applications	Installing anti-virus, anti-malware tools, and firewalls	Implementing multifactor authentication	Encrypting data traveling between the cloud and the browser	Encrypting data stored in the cloud	Intrusion detection and prevention systems and network monitoring	Perimeter security	Shielded server rooms and cages	Surveillance	Access control	Facility access logs	ICT-acceptable use policies	Password policies	User access management policies	BYOD policies	Staff training	Background checks of cloud service provider staff
Authentication issues	✓	✓			✓	✓						✓	✓	✓	✓	✓	
Denial of service attacks and botnets		✓				✓											
Use of cloud computing for criminal activities						✓											
Illegal activity by cloud service providers			✓		✓												✓
Attacks on physical security					✓		✓	✓	✓	✓	✓						
Insider abuse of access		✓	✓		✓	✓			✓	✓	✓	✓					✓
Malware	✓	✓			✓	✓									✓	✓	
Side channel attacks					✓												

Risk / Attack				
Vulnerabilities in software applications	✓	✓	✓	✓
Cryptanalysis of insecure or obsolete encryption		✓	✓	
SQL injection	✓	✓	✓	
Crime and security risks targeting cloud computing tenants				
Phishing	✓	✓	✓	✓
Domain name system attacks	✓	✓	✓	✓
Compromising the device accessing the cloud	✓	✓	✓	✓
Access management issues	✓			✓
Attacks targeting the transmission of data				
Session hijacking and session riding	✓			
Man-in-the-middle attacks	✓			
Network/packet sniffing	✓			

Arguably, a single measure can be used to address multiple threats. While the implementation of some measures resides with cloud service providers, such as physical security of data warehouses, others can be undertaken by cloud tenants themselves, particularly when they select a provider and assess the nature of the services they offer. Assessing considerations such as whether data traveling between the cloud and the browser are encrypted, whether multifactor authentication is offered, and the physical security of the data warehouse are all of critical importance.

Technical prevention measures

Technical prevention measures can be adopted by both the cloud tenant and the cloud service provider. These include patching operating systems, Internet browsers, and other software applications to protect against new vulnerabilities and malware, installing anti-virus and malware tools, and installing firewalls to protect against unauthorized access. Cloud computing providers may also implement multifactor authentication to strengthen authentication checks. Encrypting data traveling between the cloud and the browser, as well as encrypting data stored in the cloud, protects against attacks targeting the transmission of data, as well as limiting the effects of unauthorized access. Cloud service providers may also use intrusion detection and prevention systems and network monitoring.

Physical security

Cloud service providers should, arguably, provide a safe and secure data warehouse that can only be accessed by authorized personnel in order to prevent attacks against physical infrastructure as well as insider abuse of access. Physical security measures include:

- perimeter security, such as bunkers, gates and fences;
- shielded server rooms and cages that prevent eavesdropping, external scanning, and interference via electromagnetic radiation;
- surveillance, such as CCTV and security guards;
- access control, such as swipe cards, turnstiles, biometric authentication, and identity cards; and
- maintaining facility access logs.

Cloud service providers should also have effective fire management practices in place and backup power systems to prevent data loss through natural disaster or malicious attacks. While security audits should assess whether appropriate physical security measures are in place, audit rights may be beyond the scope of some cloud tenants.

Organizational policies, awareness, and training

Businesses may implement a number of organizational policies to protect against computer security threats that relate to cloud computing, as well as computer security more generally. These include ICT acceptable use policies that set out how a business's computer resources should be used, including expectations in relation to personal use, the handling of sensitive information, the installation of applications, and the forwarding of emails, which may contain malware. Password policies set out how often passwords should be changed and their complexity to strengthen authentication checks, while user access management policies set out the access rights forstaff, including that access should be discontinued when a staff member leaves an organization. Employees using their own devices in the workplace should also be subject to organizational policies, as they may be vulnerable to compromise and create more avenues for unauthorized access to cloud services.

Businesses may also provide training to staff and create awareness about computer security issues. Ensuring staff are well informed may assist in preventing social engineering attacks, such as phishing, that are not necessarily protected against by technical measures. Because of the sensitive nature of data stored by cloud service providers, they should also conduct background checks when employing staff as a preventative measure against insider abuse of access.

Service level agreements

Cloud tenants should be aware of the implications of their cloud service provider's service level agreement, which will address the issues of security, privacy, and data control. Service level agreements may also set out requirements for third-party audits of cloud service providers.

Cyber and cloud insurance

Existing cyber liability insurance holds out some limited hope of compensatingfor losses as a result of cybercrime. However, the best hope for broader coverage rests with contingent business interruption insurance adapted to the unique circumstances of cloud computing ("cloud insurance") being developed by new entrepreneurial ventures.

Crime displacement risks

Crime displacement occurs when crime moves to other locations, times, targets, methods, perpetrators, or types of offense, often as the result of crime prevention initiatives (Smith et al. 2003). Displacement concerns that relate to cloud computing may include:

- displacement to cloud service providers who do not have strong security measures;
- cloud service providers operating from jurisdictions that do not have applicable criminal provisions, have low criminal penalties, or do not have extradition treaties;
- different methods, for example, if a target is adequately protected against electronic attacks, an offender may coerce an employee through bribery or extortion; and
- displacement to perpetrators who are more highly skilled and perhaps more adept at hiding their offending activities.

Effective crime prevention requires an appreciation of these risks and the use of measures designed to address them.

Discussion and conclusion

Cloud computing extends where data may be held and expands the locations where that data may be accessed from. This may be advantageous to the authorized user, in that they can access business or personal information from different places using different devices and platforms. However, these factors also expand the scope for unauthorized access or modification of data, as well as disruption to services. In addition, cloud service providers may find that online offenders are attracted to the services that they offer and use them to carry out offenses. In the absence of empirical research into the extent of these risks, it is not known at present which are more prevalent or more important than others.

Cloud computing also prompts questions about how we think of victims, offenders, and guardians. For instance, a cloud service provider may simultaneously be a guardian, in that they are expected to protect their users' data, as well as a victim themselves, as it is their system that is being attacked. When cloud services are used in an attack, they may play a part in enabling the offense to take place, while not wittingly engaging in criminal activities. On the other hand, cloud providers may knowingly engage in, or enable their systems to be used for, criminal behavior.

However, there may also be a bright side to security in the cloud. Some cloud computing tenants that would otherwise lack the policies, procedures, and training to adequately protect their networks may actually face a reduced risk when their cloud provider manages their systems for them. At this stage, this comparison is hypothetical, as there are difficulties comparing and quantifying the associated risks and benefits.

Bibliography

Agrawal, D., El Abbadi, A. and Wang, S. (2011) "Secure Data Management in the Cloud," *Databases in Networked Information Systems* 7108: 1–15.

Aleem, A. and Sprott, C. R. (2013) "Let Me in the Cloud: Analysis of the Benefit and Risk Assessment of Cloud Platform," *Journal of Financial Crime* 20: 6–24.

Allen, J. (2010) "Data Security in a Mobile World," *GPSolo* 27: 4–6.

Aron, J. (2011) "Beware of the Botcloud," *New Scientist* 210: 24.

Banham, R. (2012) "Few Data Breaches in the Cloud – For Now," *Business Insurance* 46: 14.

Banks, L. (2010) "Cyber Criminals Using Cloud to Launch Attacks, Security Expert Warns," http://www.computerworld.com.au/article/363710/cyber_criminals_using_cloud_launch_attacks_security_expert_warns/ (accessed 3 August 2014).

Bennett, B. (2012) "Megaupload Raid Underlines Cloud Risk," *NZ Business* 26: 54.

Blanton, S. and Schiller, C. (2010) "Is There Safety in the Cloud?," *EDUCAUSE Quarterly* 32.

Blumenthal, M. S. (2011) "Is Security Lost in the Cloud?," *Communications and Strategies:* 69–86.

Cloud Security Alliance (2010) "Top Threats to Cloud Computing v1.0," https://cloudsecurityalliance.org/research/top-threats/#_downloads. (accessed 3 August 2014).

Dlodlo, N. (2011) "Legal, Privacy, Security, Access and Regulatory Issues in Cloud Computing," *Proceedings of the European Conference on Information Management & Evaluation,* 161–8.

Fortify (2010) "Vast Scale of Cloud Hacking," *International Journal of Micrographics & Optical Technology* 28: 7–8.

Grobauer, B., Walloschek, T. and Stoecker, E. (2011) "Understanding Cloud Computing Vulnerabilities," *IEEE Security & Privacy* 9: 50–7.

Hong Kong Government News (2011) *LCQ3 Information Security,* Hong Kong: Hong Kong Government News.

Hutchings, A., Smith, R. G. and James, L. (2013) "Cloud Computing for Small Business: Criminal and Security Threats and Prevention Measures," in *Trends and Issues in Crime and Criminal Justice* no. 456, Canberra, Australia: Australian Institute of Criminology.

Karagiannis, E. (2005) "Political Islam and Social Movement Theory: The Case of Hizb ut-Tahrir in Kyrgyzstan." *Religion, State and Society* 33(2): 137–50.

Knapp, K. J., Denney, G. D. and Barner, M. E. (2011) "Key Issues on Data Centre Security: An Investigation of Government Audit Reports," *Government Information Quarterly* 28: 533–41.

Kunick, J. M. (2011) "Navigate the Cloud," *Managing Intellectual Property* 210: 18.

Mell, P. and Grance, T. (2011) *The NIST Definition of Cloud Computing: Recommendations of the National Institute of Standards and Technology,* Gaithersburg: US Department of Commerce/National Institute of Standards and Technology.

Mowbray, M. (2009) "The Fog Over the Grimpen Mire: Cloud Computing and the Law," *SCRIPTed Journal of Law, Technology and Society* 6: 1–15.

Pacella, R. M. (2011) "Hacking the Cloud," *Popular Science* 278: 68–71.

Ponemon Institute (2011) *Security of Cloud Computing Providers Study,* Traverse City, MI: Ponemon Institute.

Ristenpart, T., Tromer, E., Shacham, H. et al. (2009) "Hey You, Get Off My Cloud: Exploring Information Leakage in Third-party Compute Clouds," 16th ACM Conference Computer and Communications, Chicago.

Schreiber, T. (2004) *Session Riding: A Widespread Vulnerability in Today's Web Applications,* Munchen: SecureNet.

Smith, R. G., Wolanin, N. and Worthington, G. (2003) "e-Crime Solutions and Crime Displacement," in *Trends & Issues in Crime and Criminal Justice* no. 243, Canberra, Australia: Australian Institute of Criminology.

Subashini, S. and Kavitha, V. (2011) "A Survey on Security Issues in Service Delivery Models of Cloud Computing," *Journal of Network and Computer Applications* 34: 1–11.

Trend Micro (2011) *Cloud Security Survey: Global Executive Summary,* http://newsroom.trendmicro.com/file.php/194/Global+Cloud+Survey+Exec+Summary_Final+%282%29.pdf (accessed 3 August 2014).

Wright, A. (2011) "Cloud Computing and Security," *Toledo Business Journal* 27: 20–2.

Part III

Industry, Criminal Justice, and Forensic Responses to Cybercrime

11

Hong Kong's Experience in Strengthening the Security Measures of Retail Payment Services

Shu-Pui Li

Introduction

Online financial transactions, conducted either on the Internet or through mobile channels, have gained increasing popularity in Hong Kong, as they provide convenience to the public in managing their banking accounts and payment activities. At the end of June 2014, there were around 9.1 million personal Internet banking accounts and 0.82 million business Internet banking accounts in Hong Kong. In the first half of 2014, an average of 8.3 million financial transactions, amounting to HK$363 billion, were conducted through personal Internet banking accounts per month, whereas an average of 5.8 million financial transactions, amounting to HK$5,640 billion, were conducted through business Internet banking accounts per month. With such a wide acceptance of Internet and mobile banking, it is crucial for the banking sector to implement adequate security measures.

This chapter provides an overview of the general security measures employed in online financial transactions and then gives a more in-depth analysis of the security measures that have been implemented in the electronic bill presentment and payment system, e-check and near field communication mobile payment. Finally, the chapter discusses how a new regulatory framework to be implemented in Hong Kong will help ensure the safety of retail payment instruments.

General security challenges and measures

User authentication

Challenges

A major challenge associated with online banking is the need to ascertain the identity of users before allowing them to access their accounts and conduct financial transactions. Unlike conducting financial transactions at a bank branch at which the bank staff can ask a customer to provide an identity card or passport for the purpose of user authentication, it is impossible to do so online.

Security measure: Adoption of two-factor authentication

Two-factor authentication requires a customer to prove his or her identity through the use of two independent factors. Given that passwords are commonly used as the basic factor of authentication, banks are required to adopt a second factor, which cannot be easily stolen by fraudsters, for customer authentication before allowing customers to conduct high-risk online financial transactions. Examples of second-factor authentication include digital certificates, one-time passwords generated by a security token or through Short Message Service (SMS) through the mobile network. High-risk online financial transactions include unregistered third-party fund transfers and payments, and change requests concerning customers' sensitive information. Nonetheless, such transactions should not be allowed to be conducted through mobile banking as the same device may be used for receiving SMS-based one-time password, which will inevitably defeat the purpose of two-factor authentication.

Unauthorized access of payment card details

Challenge

Due to the growth of e-commerce, some payment gateway service providers have offered payment solutions to facilitate online payments at shopping websites. The most commonly used payment instrument for this purpose is a credit card. Some transactions require only general credit card information to effect a payment, such as the credit card number, the cardholder's name, and the expiry date of the credit card, all of which are printed on a physical card. The main threat arising is the possibility of unauthorized access and use of credit card information by a fraudster.

Security measure: Fraud detection mechanism

To protect credit cardholders during the online payment process, card issuers have implemented various fraud detection systems to alert cardholders to any suspicious transactions. By analyzing the payment pattern of a particular customer, such as merchant types and transaction amounts usually associated with the customer, card issuers can screen suspicious transactions and notify a cardholder regarding those transactions. In addition, some card issuers have implemented additional security measures in conjunction with payment card associations (e.g., VISA and MasterCard) through which the cardholders may be required to input an additional password and answer some other security questions before effecting an online payment.

Attempts to steal user IDs and passwords

Challenge

Internet banking and mobile banking services face certain technological risks. Some fraud cases involve fake websites seeking to attract bank customers to divulge confidential personal information. Other cases may involve Trojan software and other highly infectious computer viruses, which may be installed when a user clicks on hyperlinks embedded in emails or browses an infected website with pop-up advertisements. Once planted in a computer, the Trojan software may be activated when the user accesses certain websites: it can then capture the keystrokes of the infected computer's user, which could in turn lead to leakage of sensitive personal information such as user IDs and passwords.

Security measure: Public education

In addition to the initiatives of the Hong Kong Monetary Authority (HKMA) and banking sector, members of the public should also take part in upholding the security of online transactions conducted through Internet and mobile platforms. The HKMA has from time to time reminded the public to remain vigilant when using online banking services. For instance, they are reminded to avoid accessing bank websites through hyperlinks provided in emails, Internet search engines, or suspicious pop-up windows. They should also be careful about opening suspicious emails with attachments from senders they do not recognize. Most importantly, they should never give sensitive information to a third party. Whenever there are fraudulent emails or websites purporting to be related to a bank, the HKMA will immediately issue a press statement and alert the public to the potential fraud.

Failure of payment service providers to identify risks and implement appropriate security measures

Challenge

While technological advances provide an opportunity to the market to develop innovative payment services, they may also create new types of risks that cannot be easily identified in the absence of thorough study and assessment. Failure to identify such risks will render it impossible for payment service providers to implement appropriate security measures, and thus undermine users' confidence if such hidden risks surface and lead to significant damage.

Security measure: Pre-launch audit and self-assessment

To ensure that security measures have been adequately and properly implemented for a new payment service which involves substantial enhancements to the computer system, it is the practice of Hong Kong Interbank Clearing Limited (HKICL), the clearing house for interbank payment and settlement in Hong Kong, to engage an audit firm to conduct pre-launch audit of such services. The pre-launch audit covers, among others, an independent assessment of the security controls that have been put in place. It is upon the positive confirmation of the audit firm that the payment service will be launched. Similarly, before rolling out a new service, participating banks are also required to conduct self-assessment and confirm with the HKMA that they have complied with the relevant regulatory requirements.

Security measures for retail payment initiatives

This section explores some specific security measures that have been implemented for electronic bill presentment and payment systems, e-check, and near field communication mobile payment services.

Electronic bill presentment and payment (EBPP) system

Background

Launched in December 2013, the EBPP system provides a consolidated channel for bill owners to receive, manage, and pay e-bills through the Internet banking platform. It is operated by HKICL, which is equally owned by the HKMA and the Hong Kong Association of Banks. It supports a wide range of transactions including 1) business-to-customer billing and payment, 2) business-to-business billing and payment; 3) cross-border billing and payment; and 4) charity payments and presentment of e-donation receipts. At the end of December 2014, 19 banks and 112 merchants covering schools, property management companies, telecommunication operators, and charity organizations have joined the system.

It is easy for bill owners to use and enjoy the benefits of the EBPP service. It takes only three simple steps: 1) one-off subscription of e-bill for a particular merchant; 2) receipt of e-bill summary; and 3) making e-bill payment (please refer to Figure 11.1).

As it is necessary to standardize the information of e-bills that are shown on the Internet banking platform, only the basic information, such as merchant name, issue date, bill account number, bill amount, and payment due date will be provided by the merchant. To facilitate bill owners' ability to view e-bill details which are maintained on merchants' websites, HKICL has developed a single sign-on (SSO) platform under which the bill owners can access e-bill details directly on the Internet banking platform without the need to log in to the merchant's website separately (please refer to Figure 11.2).

Security measures

In view of the vast amount of sensitive information and the large number of merchants involved, the EBPP system has implemented a number

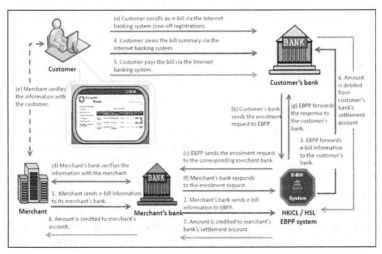

Figure 11.1   High-level processing flow of EBPP

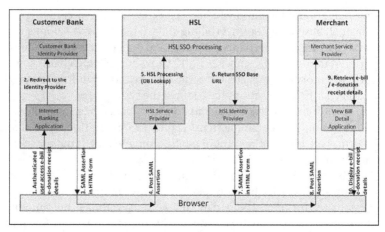

Figure 11.2   Technical perspective of the SSO mechanism

of security measures to avoid data leakage and to ascertain the identity of bank customers before allowing them to make bill payments. Such security measures are, briefly, as follows.

Two-factor authentication

Since 2005, the HKMA has required all banks to implement two-factor authentication for third-party funds transfers which are classified as high-risk transactions. Such a requirement also applies to bill payments, which are usually destined for merchants (except the government and utility companies, which generally entail lower risk). Basically, a customer needs a second factor which is in his or her physical possession for identity verification in addition to the normal login ID and password. Three methods commonly adopted in Hong Kong are digital certificates, security token-based one-time passwords, and SMS-based one-time passwords (OTP).

To ensure continuing effectiveness of two-factor authentication, the HKMA has closely monitored market development and requested the banking sector to strengthen the security controls where the situation warrants. In view of the heightening threat arising from technological advances, the HKMA issued a circular in 2009 to lay down some guiding principles for banks to use in addressing such risk. For instance, if an OTP is used for an online high-risk transaction, banks should notify their customers immediately after the transaction via an effective alternative channel. In addition, they should ensure that the digital certificate and its associated private key is non-duplicable and is stored in secured media (see HKMA 2009). Such measures have proved effective in mitigating the risks of unauthorized payments and will continue to be adopted for all high-risk transactions including the EBPP service.

Secured data transmission network
All payment files for small value transactions, such as auto-debit, credit transfer items, debit card, and credit card transactions are exchanged between banks and HKICL through a secure data transmission network, namely ICLNet, for interbank clearing and settlement. It is a secure, scalable, and high-performance private Internet protocol-based network which connects the respective computer systems of the financial institutions and other licensed financial entities in Hong Kong to exchange electronic data safely and efficiently. Following the existing arrangements, participating banks of the EBPP service are required to exchange payment files and data files containing e-bill summaries with HKICL through ICLNet to safeguard data security.

Authentication of SSO platform
The SSO platform is an Internet-based real-time service operated by HKICL. It serves as a connecting bridge between the Internet banking platform and the merchant's website and performs authentication checking before allowing bill owners to view e-bill details directly on the Internet banking platform. It operates on the industry standard known as Security Assertion Markup Language (SAML) 2.0 and is equipped with strong security measures including the adoption of digital signing, encryption, and secure network transmission through Hypertext Transfer Protocol Secure (HTTPS). It has a well-proven track record and is difficult, if not impossible, to crack using the current technology. Due to the stringent security measures, merchants can be assured of the identity and authenticity of the customers before they can access the e-bill details through the SSO platform.

E-check

Background
Despite the availability of a large variety of electronic payment instruments, checks continue to be heavily used by the general public to conduct a wide range of transactions. According to the statistics of HKICL, around 500,000 checks are cleared through HKICL per day. Together with "on-us" checks which are processed by the banks without the knowledge of HKICL, the number of checks handled by the entire banking sector per day could be much higher. While checks have proved effective in some transactions, they require certain degrees of manual operations in the issuance, delivery, and presentment processes. In order to enhance the

operational efficiency of check payments, it is desirable for the society to introduce a new payment instrument, namely e-check, which mimics the key features of the paper check on the one hand and enables it to operate through a highly automated and secure platform on the other hand.

Scheduled for launch in December 2015, e-checks will be signed, issued, delivered, and presented through electronic means. The operating model is briefly described as follows (see Figures 11.3 and 11.4).

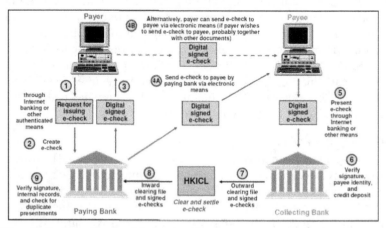

Figure 11.3   Operating model of e-check (presentment of e-check through Internet banking platform)

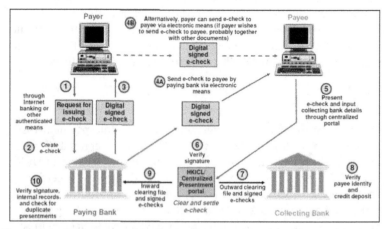

Figure 11.4   Operating model of e-check (presentment of e-check through centralized presentment portal)

Stage 1: Issuance of e-check. Stage 1 involves the payer giving payment instructions including payee name, check date, and check amount to his or her paying bank through the Internet banking platform or other authenticated means. Since issuing e-checks allows the payer to transfer funds to a third party and thus constitutes a high-risk transaction, two-factor authentication is needed for issuing e-checks. Upon confirmation of the identity of the payer, the paying bank will generate an e-check which is in Portable Document Format (PDF). It will carry the digital signatures of the payer and the paying bank. For the purpose of signing e-checks, the payer can either ask the paying bank to apply, renew, and keep custody of a digital certificate on his or her behalf or directly apply for a digital certificate from one of the recognized certification authorities under the Electronic Transactions Ordinance.

Stage 2: Delivery of e-check. The payer can ask the paying bank to send the e-check to the payee directly after the generation of the e-check. Alternatively, the payer can download the e-check and send to the payee through electronic means, such as a secured email system.

Stage 3: Presentation of e-check. Upon receipt of e-check, the payee can present it either through the Internet banking platform or the centralized presentment portal operated by HKICL.

Where an e-check is presented through the Internet banking platform, the collecting bank will 1) verify the genuineness of e-check, 2) verify if the payee's name matches with the bank account holder's name, 3) check with HKICL to ascertain if the e-check has been presented or settled earlier, and 4) post the transaction to the payee's account.

Where an e-check is presented through the centralized presentment portal, the payee has to provide the bank's name and bank account number for the deposit of the e-check. HKICL will 1) verify the genuineness of the e-check, 2) ascertain if the e-check has been presented or settled earlier, and 3) pass the e-check information to the collecting bank for further processing.

Stage 4: Interbank clearing and settlement. Similar to the interbank clearing and settlement process for paper checks, HKICL will submit the presented e-checks to the paying banks to check if there are sufficient funds in payers' bank accounts to honor the payment obligations.

Major security challenges

Although the term "e-check" has been widely used to refer to a check-like electronic payment instrument in some jurisdictions, it involves the use of paper in the issuance process in most cases. For instance, "e-check" in the US, which is also known as remote check capture, requires a payer to issue and deliver the paper check to a payee. Upon receipt of the paper check physically, the payee is allowed to take a scanned image of the paper check and present the electronic image to his or her bank through either a mobile application or over the Internet instead of depositing the check at a bank branch.

In contrast, the proposed e-check model in Hong Kong is an "end-to-end" electronic payment instrument, without the use of paper or the need of physical delivery and presentment throughout the process. Based on our research, it is probably the first of its kind around the world. Being the first place to launch the e-check initiative also means that Hong Kong does not have the benefit of learning from the experience of others. To date we have encountered three major security challenges in the implementation of e-checks, which are briefly described as follows.

Forgery and alteration of content.   To most people, e-checks are simply a PDF file containing the essential information of a paper check and can be created and edited in a way similar to other electronic files. This leads to a question as to how the public can easily and effectively verify the genuineness and integrity of the e-check.

Duplication and multiple presentations.   By its nature, an e-check can be easily replicated, and it is not possible for someone to tell the difference between the original electronic record and the electronic copy. This leads to a question as to whether the e-check infrastructure is able to restrict the same e-check from being presented multiple times.

Data leakage.   Since e-checks can be delivered by a payer to a payee over the Internet, and it is possible for both the payer and payee to keep the electronic record of e-checks in their personal computers or mobile devices. This leads to a question as to whether there are effective data protection measures to guard against possible data leakage.

Security measures

It is important to strike a balance between the adequacy of the security measures of e-checks and the user-friendliness and convenience of use. While the above challenges may be addressed by modifying the e-check infrastructure from an open platform to a closed-loop platform (i.e., no one can access the e-check unless he or she has joined the same platform and the e-check cannot be replicated outside the platform), it will inevitably undermine the user-friendliness, and thus the adoption, of the e-check system. Thus, the closed-loop platform will not be considered at this stage. Instead, we have implemented the following security measures for e-checks to address such challenges.

Digital signature using public key infrastructure (PKI) technology. Each e-check will be digitally signed by both the payer and the paying bank using PKI technology. Basically, PKI technology involves the use of a pair of private keys and public keys in the encryption and decryption processes. Taking the digital signature of the paying bank as an example, it can only be generated by the paying bank which possesses the private key of such digital signature. The validity of the digital signature shown on the face of the e-check can be verified by anyone using the corresponding public key of the digital signature, which is a piece of information that is made available to the public by the recognized certification authorities.

While the operation of PKI technology with e-checks may appear somewhat technical to the general public, its benefits are easy to understand. First, based on the current technology, it is impossible for any fraudsters to create a fake digital signature of the paying bank, and thus a fake e-check. Secondly, it is impossible for anyone to tamper with the e-check once it is digitally signed by the paying bank. Any attempt to alter an e-check by, for example, amending the payee name and check amount, will cause it to become invalid and not presentable to any bank.

Virtual checkbook. It is not necessary for the paying bank to print and mail any physical checkbook to the payer for the purpose of e-check. The virtual checkbook is kept by the paying bank and can only be accessed by the payer when he or she successfully logs on to the Internet banking platform. This eliminates the risk of loss of the physical checkbook in transit or any unauthorized access.

Maintenance of e-check issuance records by the paying bank. Since e-checks are issued by the payer on the Internet banking platform, the paying bank will be notified once the e-check has been issued, thus providing an additional channel for the paying bank to verify the information of the deposited e-check.

Implementation of centralized presentment portal. Regardless of the presentment channels chosen by the payees, all deposited e-checks will be ultimately sent to the centralized presentment portal for duplicate presentment checking. Such a portal maintains a database of the e-checks that have been presented to the collecting banks and the e-checks that have been settled by the paying banks. Any e-checks that have been presented or settled earlier will be rejected.

Adoption of secured delivery channels. To avoid any unauthorized persons reading e-check information, the payer can encrypt an e-check before sending it to the payee's email account. In addition, such delivery should be made through a secure email system to further reduce the chance of data leakage.

Based on the advice of independent security experts, the above measures are considered effective in addressing the security challenges that have been mentioned before. The HKMA will conduct consumer education programmes to remind the public of some smart tips to use e-checks safely and efficiently while monitoring the effectiveness of such security measures on an ongoing basis.

Near field communication (NFC) mobile payment

Background

NFC mobile payment is a contactless retail payment channel for the use of NFC-enabled mobile phones. A user can make contactless payments at retail outlets by tapping his or her mobile phone on a contactless reader installed at retail outlets. With the popularity of smartphones and the widespread use of contactless card payment in Hong Kong, the new payment channel has recently drawn the market's interest.

To facilitate the development of NFC mobile payment in Hong Kong, the HKMA appointed a consultant to conduct a study on NFC mobile payment infrastructure development in Hong Kong and released the study results in March 2013. One of the recommendations of the study was to establish a common set of standards and guidelines on the implementation of NFC mobile payment solutions in Hong Kong. In this context, the HKMA coordinated with the Hong Kong Association of Banks to develop a set of standards and guidelines ("Best Practice") for the industry to follow. The Best Practice Guidelines, issued in November 2013, comprise technical standards, security requirements, and operational processes (HKMA 2013).

As of December 2014, seven banks have issued NFC mobile credit cards in collaboration with two card associations. A stored-value card operator has also launched an NFC mobile stored-value card which can be used in both the public transport systems and retail shops.

Security measures

Similar to personal computers, mobile devices are also susceptible to various security attacks. However, it is not common for users to install anti-virus software and regularly update the latest virus list in their mobile phones as in the case of desktop computers. This will render mobile devices more vulnerable to the risk of being hacked. To tackle the issue, several security measures have been implemented for NFC mobile payment.

Use of tamper-proof device for the storage of payment card credentials. To protect payment card credentials from being counterfeited or tampered with by any unauthorized parties, the card credentials are separated from mobile phone storage where phone book contacts, short messages, photos, and other mobile applications are stored. Instead, card credentials are kept in a tamper-proof chip, commonly known as a secure element. The security level of the chip is the same as the banks' chip-based ATM cards. Retrieval of data from the chip is restricted to go through the chip's interface which is safeguarded with security controls. Only preapproved mobile applications, such as the payment service provider's mobile application, are allowed to access the chip. In addition, the data for transmission are encrypted by the chip to protect the confidentiality of the data.

In the case that payment card credentials from different payment service providers are stored on the chip, they should be adequately segregated from each other within the secure element. This segregation is managed through the creation of the proper secure domain structure and hierarchy. Each payment service provider owns its security domain where the credentials are stored.

Payment card authentication
When a mobile payment card is tapped on a contactless reader, it is authenticated by the reader before a payment transaction can be completed to protect against counterfeit cards. The card authentication process involves the transmission of payment card data (such as card number, expiry date) and a cryptogram to the reader. The cryptogram is generated from a cryptographic key residing in a secure element. The key itself is never transmitted outside the secure element. To avoid the transaction data being captured and reproduced during a payment transaction, each transaction generates a unique cryptogram along with the transaction data.

Cardholder verification. Contactless payment allows a user to simply tap her mobile phone on a contactless reader to complete a payment transaction. For transactions with values below a threshold, no handwritten signature or PIN code is needed to verify the cardholder. To protect against lost and stolen mobile phones, mobile credit card service providers require a user to authenticate his or her identity before making a payment or enquire about the transaction history. The cardholder verification process usually requires a password to activate the mobile wallet. An unauthorized person taking possession of another person's mobile phone will not be able to initiate a payment unless he knows the password of the mobile wallet.

Recent technological developments have enabled the use of biometric processes, such as fingerprint, face, and voice recognition, for cardholder verification. The uniqueness of biometric data ensures that the person attempting to make a transaction is indeed the cardholder of that mobile payment card. One concern with using biometric authentication for payment transactions is the privacy of the personal information. To address the concern, hashing algorithms should be applied to convert the biometric data to a hash value before it is stored in a mobile phone. The algorithms are designed such that a person cannot restore the biometric data from the hash value.

Enquiry about transaction history. A mobile payment card user can enquire about the transaction history through the mobile payment application installed in his mobile phone. The user can check if any unknown transactions have been made and if the transaction values are correct, without waiting for a monthly statement or browsing the transaction history through the Internet banking channel. The feature shortens the elapsed time to detect any fraudulent transactions and thus reduces the chance of fraud cases.

Risk mitigation measures. As NFC mobile payment is relatively new to the general public, risk mitigation measures have been put in place to limit risk exposure. The mobile credit card service providers generally agree to adopt a prudent approach in order to get users familiar with this payment method and build up their confidence.

Limit on transaction values. At the moment, a limit on transaction values, generally at HK$500 per transaction, has been imposed on mobile credit cards. The determination of the limit is based on the risk appetite of cardholders and is subject to review as necessary. For mobile stored-value cards, the ceiling is the outstanding balance in the cards.

Restrictions on remote payment with NFC mobile payment cards. To further reduce the risk of a user, NFC mobile credit cards issued in the market are restricted to point-of-sale contactless payments through contactless readers. The cards are not allowed to make remote payments, commonly known as card-not-present payments. In remote payment scenario, cardholders and merchants are not face to face. The risk of fraudulent transactions is generally higher than the point-of-sales payment where a physical presence of the cardholder is needed to make a purchase. The restriction helps prevent fraudsters' attempts to make online purchase with the card credentials obtained from compromised mobile payment cards.

The way forward

As NFC mobile payment is in an evolving stage, new technologies may be adopted and innovative services may be launched. To cope with the fast-changing situation, the adequacy of the security measures has to be reviewed regularly. In addition, as the general public is cautious regarding the perceived risks associated with mobile payment, the risk exposure of any mobile payment services should be assessed and limited to maintain the public confidence.

Regulatory regime for stored-value facilities and retail payment systems

The overall design of the new regulatory regime

Currently, the regulatory framework for stored-value facilities (SVF) in Hong Kong only covers multipurpose stored-value cards. Meanwhile, the payment industry adopts an informal self-regulatory approach to ensure the safety and efficiency of selected retail payment systems (RPS) that are commonly used.

The HKMA and the Financial Services and the Treasury Bureau (FSTB) have recently proposed to amend the Clearing and Settlement Systems Ordinance to establish a regulatory framework for multipurpose SVF (which can be either device-based, or non-device-based, such as those using network accounts) and RPS. The main aim is to ensure that these SVF and RPS are safe and sound and that the float money stored in the SVF is properly protected.

Specifically, the issuers of the SVF which are subject to regulation would need to obtain a licence from the HKMA under a mandatory licensing regime. Their businesses need to meet certain criteria in order to be licensed. For instance, they need to have physical presence in Hong Kong. Their principal business must be the issuing of SVF. In terms of financial strength, they must meet the minimum ongoing capital requirement of HK$25 million. Importantly, the issuer must have adequate risk management policies and procedures for managing the float so that there will be sufficient funds for the redemption of outstanding stored value.

As regards RPS, the HKMA would have the power to designate certain RPS if any disruption to their operations could have systemic concerns.

Technology risk management: A key part of the new regulatory regime

As far as the regulated SVF are concerned, one of the licensing criteria is that the issuer must have in place appropriate risk management processes and measures. They may include, for instance, security and internal controls to ensure the safety and integrity of data and systems, effective fraud monitoring and detection measures, and robust and well-tested contingency arrangements to address operational disruptions. The HKMA would be empowered to conduct examinations of the licensees' activities to ensure that SVF licensees will continuously meet the statutory requirements including the licensing requirements such as those regarding technology risk management.

As for the RPS, the HKMA would also be empowered to oversee their operations (including their technology risk management aspects) by, say, gathering information, giving directions, imposing operating rules, making regulations, issuing guidelines, and so on.

At the time of writing, an amendment bill has been tabled in the Legislative Council and it is expected that a Bills Committee will be formed in 2015 to scrutinize the bill. It is expected that the new legislation will be enacted in 2015 or 2016 and, after a transitional period of one year, the new regime will become effective.

Conclusion

Retail payment technologies are evolving fast and, by making day-to-day payments more convenient and secure, are set to bring enormous benefits to the general public. Nonetheless, innovations also bring new technological risks. The HKMA will continue to monitor market trends and put in place appropriate measures to ensure that the public can enjoy cutting-edge payment technologies and that any payment instruments which are widely used and could have systemic implications are safe and sound.

Bibliography

Hong Kong Monetary Authority (2009) "Strengthening Security Controls for Internet Banking Services," 13 July, revised 1 August 2011, http://www.hkma.gov.hk/eng/key-information/press-releases/2009/20090713-3.shtml (accessed 16 February 2015).

Hong Kong Monetary Authority (2013) "HKAB Issues Best Practice on NFC Mobile Payment in Hong Kong," 25 November, http://www.hkma.gov.hk/eng/key-information/press-releases/2013/20131125-3.shtml (accessed 16 February 2015).

12
Banking Security: A Hong Kong Perspective

Paul Tak Shing Liu

Introduction

At present, financial institutions use many electronic systems for the operation of their financial environment. In Hong Kong, many bank host systems, for the processing of daily financial transactions, use IBM systems, COBOL, and relational databases, while others use Unix, C/C++, and relational databases such as Oracle. An extensive array of other electronic systems are built around the bank host system. These include branch teller systems, internal users interfacing systems, ATMC, ATMP, JETCO gateway, SWIFT, CHATS, systems for printing reports and statements, systems for generating electronic statements, phone banking systems, Internet banking systems, mobile banking systems, anti-money laundering systems, electronic workflow systems, inward and outward clearing systems, signature card systems, check imaging systems, forex trading systems, securities trading systems, brokerage systems, stock quotation systems, insurance policy management systems, insurance broker systems, and others.

Security controls in banking in Hong Kong

Traditionally, the banking industry uses very strict security controls over all operational and transaction-related procedures. In terms of segregation of duties, all systems administration must have dual login controls and very rare network protocols for host banking system communications. Networks employ multiple serial firewalls along with many other network security devices. All desktop PCs are configured to eliminate access through CD-ROM, DVD-ROM, USB, Bluetooth, Wi-Fi, and floppy discs. Notebooks and mobile devices are subject to very restrictive controls and are limited to very senior management staff only. Bring-your-own-device (BYOD) is prohibited throughout the bank's network.

All PCs are deployed with anti-virus applications, and all Intel servers are installed with different brands of anti-virus applications. Email servers have yet another different anti-virus application installed. SMTP servers are also installed with a different brand of anti-virus application. Proxy servers are used for Internet access by internal users only. Reverse proxy servers are used for Internet transaction-related web applications (e.g., eMPF). Most of the workflow applications are designed using browser-based client formats so that no one can directly download all data. In addition, these workflow applications can easily be controlled using firewalls, web application firewalls, and packet filtering. Internet banking itself also uses multiple security control mechanisms including multiple serial firewall configuration, user ID-password, SMS as a second factor, eCert as a second factor, page sequencing control, and cookies.

Security control risks in banking in Hong Kong

Security controls in banking historically raise a number of complex issues. The computing design language of the host banking system and network communication protocols tend to be old-fashioned, making it difficult to hack into systems from the Internet. Most likely, host systems will be surrounded by multiple firewalls, IPS, packet filtering, and other measures. Nonetheless, a number of security risks remain if not properly managed.

Insider risks

As few application developers know the internal operations of the bank host, it is most likely that hacking will be committed by internal staff. Although quite costly, one way to prevent this is to set up four different technical teams. The original team will be responsible for the actual application development and maintenance. Another team will do the source codes audit. The remaining team will handle a production test of any modifications. Then, finally, the executable will be released and deployed to production by another deployment team. From an operational risk point of view, this results in a very secure and reliable system since the source code audit team and production test team will conduct their jobs logically in parallel with the design and development team. The deployment team will only be able to handle the deployment of modification systems without access to the source codes directly.

Developer risks

Another problem concerns firefighting, or emergency debugging of the bank host system. Most of the time, whenever a host banking developer accesses the production, no one knows precisely what the developer is doing. Therefore, it is necessary to enforce the above system deployment procedure. The firefighting developer should gather problem scenario information and log data for analysis, debugging, and testing on an identical testing environment. Then all modifications, audit, and testing should be performed. Finally, the modification should then be deployed. This may take a much longer time, but it will be a lot more secure than simply allowing developers to access the production directly. In practice, however, not all financial institutions would take this approach owing to the high costs involved. Testing is an important part of application development. Beside functionality tests, system integration tests, user acceptance tests, volume tests, stress tests, random tests, and testing of security loopholes become very important. Such testing is part of security control even before the deployment of any application. Although we have many security control devices and systems, a properly designed application will definitely reduce the possibility of it being hacked.

Operating system risks

The Windows operating system, Microsoft products, and related products are a lot more vulnerable to hacking than any other operating system, such as Linux. However, due to the extensive use of Debian Linux in many mobile devices (MacBook, iPhone, iPad, Android, etc.), it is necessary to properly protect such devices. In Windows, we do have personal firewalls that can control data traffic from and to individual applications beside the control of IP addresses and ports. In Debian Linux devices, there should be similar proxy filtering and firewall applications in order to control data nature and throughput. This should give users the power to control the flow of data both inward and outward through the network. Therefore, BYOD can be a problem if the device is not properly installed and protected. Simple installation and configuration will always improve security levels for such devices used within organizations.

Data access risks

In a highly competitive banking environment like Hong Kong, it is necessary to control the installation of and access to applications on desktop/notebook PCs. Access to data via USBs, CDRs, and DVDs with read or write access also needs to be controlled. Similarly, Bluetooth access and Wi-Fi network access using encryption may be needed. Desktop system setups must be erasable and installable. Most data should be stored on banking data/file servers for both security and backup purposes. This should protect the bank's interests where PCs or devices are stolen.

User access to banking information should be via browser-based systems that can easily control who can access the system and its data. Such systems can slow down data access and can prevent large amounts of data from being leaked. Browser-based systems give everyone the option of accessing data and information any time anywhere, but many institutions still prefer storing data and files on desktop PCs for reasons of personal convenience.

Signature risks

In workflow systems, there are two different definitions of electronic documents – 1) imaged documents and 2) electronic documents decomposed into meaningful data fragments. Taking an image of a physically signed document is very different from a digitally signed document using an electronic certificate. Signing an electronic document using an electronic certificate that produces a digital signature is much more secure and universal than simply taking a photograph of a physically signed document. Digital signatures using electronic certificates are even better than signing a personal physical signature on an image capturing device, such as those used by FedEx or UPS. In the latter case, if the digitally captured physical signatures are not properly protected, hackers can easily forge the other person's identity using the captured data stream of the physical signature.

When we compare the use of passwords with the use of electronic certificates, it is clear that electronic certificates are considerably more advanced mathematically. However, in terms of security, this is not necessarily the case, as passwords can be protected by hashing techniques. Password comparisons can be performed without direct access to clear text while electronic certificates are more secure in the sense that digital signatures generated from electronic certificates cannot be easily reproduced simply with knowledge of the public key of the electronic certificate. However, if the renewal process of electronic certificates is not secure enough, the security of electronic certificates can be much worse than for passwords. For example, Digi-Sign's file-based electronic certificate renewal process can be easily compromised if the desktop PC concerned is compromised. No matter how secure the banking process would be, it could be jeopardized without Digi-Sign's knowledge.

Named electronic certificates and public-private key risks

Named electronic certificates (or named public-private keys) have been in existence for many years and are endorsed by legislation in Hong Kong. They have been in use for many years in Hong Kong at a cost of around HK$200. Unfortunately, the renewal process employs a file-based process which is degraded compared with the security criteria in the banking environment. Although storing electronic certificates and signing electronic documents use Java-based USB tokens, USB token drivers usually have no update in new Windows operating systems and cannot be used in Linux operating systems. Also, named electronic certificates cannot perform those operations described in the design of electronic cash. All of these considerations make the use of named electronic certificates problematic.

Nonetheless, electronic certificates and digital signatures are excellent security mechanisms for use on insecure communication channels. Using a recipient's public key to encrypt data and files, or using a sender's private key to digitally sign electronic documents for intended recipients through unencrypted communication channels or via the Internet is the original intended use of public-private key pairs. Such data or electronic document transmission can be much more difficult to implement using user IDs and passwords.

Anonymous electronic certificates

Anonymous electronic certificates, on the other hand, can perform all of the operations offered by named electronic certificates and more. They can be free of charge, and single individuals can apply more than one anonymous electronic certificate for different purposes. Anonymous electronic certificates can be used to perform all unnamed operations, such as electronic cash. Therefore, anonymous electronic certificates should dominate the individual market in the future. The only difficulty that arises at present concerns the legality and binding nature of anonymous electronic certificates.

Public-private key infrastructure (PKI) is used in a variety of recent electronic banking platforms that raise a number of associated security concerns.

Electronic cash

Electronic cash, "eCash," requires the operational algorithms of a public-private key pair. Electronic cash also needs many anonymous electronic certificates to perform normal life financial transactions without being identified. Therefore, an anonymous electronic certificate is important for all electronic cash operations. Electronic cash is similar to normal paper money currency and requires certain regulations to maintain a reasonable supply so that inflation can be properly controlled. Governments should control and issue electronic cash as authorized currency with a limited total volume available in the market. Settlement of eCash transactions should be allowed in accordance with normal procedures for currency.

Electronic checks

Electronic checks (cheques), ("eChecks") have been used in Hong Kong since 2013. Digital signatures by payer on electronic checks are one of the requirements in order for the bank to know that the information on the check is correct. However, in the case of electronic checks, the issuing bank has the full name of the payee which has been entered by the payer directly. Therefore, when the electronic check is presented to any other bank, the issuing bank can always verify the payee's name on the check. As a result, the payer's digital signature is not employed to protect the payee and is mainly used to protect the payer in the case of third-party account transfers. Ideally, electronic checks should be encrypted using the payee's public key to protect the payee's interest in that check, as it is possible to find another individual or company globally with an identical name, especially if banks in some countries allow BVI accounts to be opened indiscriminately.

Near field communications for mobile payments

NFC (near field communication) has been deployed for mobile payments using credit cards for some time now. However, many people claim that credit card information can be easily captured by other devices because the credit card terminal has no secure identification of the credit card itself, and the credit card itself has no secure mechanism to identify the credit card terminal. As a result, credit cards can easily leak card information through other touchless devices with NFC capabilities. Further difficulties arise if the credit card transaction is not encrypted on either side.

To address these issues, credit card terminals must be able to generate their own identity using server-side private keys to encrypt certain time-dependable information. This identity, along with the encrypted transaction details, should be decrypted by the application on the credit card itself to identify the credit card terminal. The credit card should also encrypt a sequence of information (card number, phone specific data, time stamp, time-dependable codes, and transaction details from specific credit card terminal) before sending that encrypted information back to the credit card terminal for transaction processing. This would definitely reduce many intended and unintended hacking problems. Simply sending out clear text of credit card numbers and related information is irresponsible.

Electronic bill presentment

Electronic bill presentment and electronic bill payment have been in operation since 2013. Using PKI (public-private key infrastructure) and SOA (service-oriented architecture), electronic bill presentment can be easily accomplished. There is no need to establish a completely different infrastructure to handle electronic bill presentment as any user of Internet banking can digitally sign a form to give the bank the power of attorney to extract all the bills belonging to the user from all other institutions. This means that extra registration of bills for individual users and creating an accompanying large infrastructure is unnecessary.

Security risks associated with other banking channels and procedures

Phone banking

Phone banking is an important channel for retail banking transactions. Accessing phone banking requires numeric user ID and password owing to the restricted keypad on phones. Normally, this kind of security mechanism is not a problem; however, with the appearance of VoIP (Voice over IP), it has become relatively easy to hack this information online. Accordingly, phone banking security controls must be extensively reviewed to counter such risks.

Internet banking

Internet banking has been developed and implemented for more than ten years with extensive usage by young banking customers. Mobile banking and mobile securities trading have similar security features to Internet banking and rely on logins using user IDs and passwords to identify users. General "https" or port 443 use published public keys to encrypt the data sent to specific servers. Public key and Diffie-Hellman are used for 3DES encryption key exchange. Page sequencing will narrow down the security window to per page control and second factor authentication will improve the security of third-party account transfers. This dramatically reduces the possibility of individual accounts being hacked.

Second factor authentication is used for both login and third-party account transfers and increases the difficulty of being hacked during login and third-party account transfer authorization. Digital signatures are a second factor generated from the private key of the electronic certificate making hacking and middle-man attacks impossible. The only problem with electronic certificates is that the electronic certificate renewal process security standard is not compatible with banking standards. In addition, the cost of issuing named electronic certificates can be a problem which can be resolved by using anonymous electronic certificates. SMS as a second factor adds another channel of authorization so that middle-man attack becomes very difficult. The only problem is that parts of SMS messaging channels are not totally secure. Synchronized temporary passcode as a second factor generated from independent offline devices on the customer side of transactions is an excellent way in which to tackle middle-man attacks. The only problem is that the cost of distributing new devices every five or six years can be substantial.

Wealth management scoring

Wealth management scoring for personal loan applications and credit card applications can be both very useful and very dangerous. Scoring itself is an approximate analysis of individual's risk characteristics and tendencies regarding investment choices, financial management styles, and repayment tendencies. It is only approximate and has limited accuracy. However, since 2008, HKMA (Hong Kong Monetary Authority) has required scoring to be accurate and precise, which is practically impossible, thus creating a very high operational risk on banking institutions. Scoring itself has predetermined parameters which are derived from customers' behavioral data based on historical transactions which can never be treated as accurate and precise.

Since scoring may affect the quality of a bank's asset portfolio, it is very important to protect the scoring systems used in all related applications. Both internal and external hackers may modify scoring mechanisms and results to undertake illegal activities, resulting in the financial position of the affected financial institution being jeopardized.

Segregation of duties

Segregation of duties is the principal procedure required in banking operations, especially in traditional manual operations or transactions. It must also be implemented both in electronic systems and physical manual procedures to reduce possible risk exposure of operation and security control. In order to implement segregation of duties in electronic operations, it is necessary to find out at least two different ways of computing the final results without going through the same system twice. This provides protection from possible operational risk exposure. General hackers would find it very difficult to hack into different systems in respect of the same operation or transaction. Therefore, segregation of duties offers not only protection from general operational errors but also protection from both internal and external hackers.

Address verification

Customer addresses and customer email addresses are becoming blurred when electronic statements are widely used. Since proof of address is still very important in cases of public voting rights, opening bank accounts, and other operations, any bank or financial institution must have mechanisms to verify customer addresses regularly. For example, a bank must ask customers to verify their address by mailing a verification request form to customers regularly every two years. If a bank posts customer statements regularly to customers' email addresses, then there is a risk that hackers may capture and modify statements in order to route customer information to fabricated addresses.

Cloud computing risks

Cloud computing involving the public use of databases, memory, and hard disk storage provides ready accessibility to data. Unfortunately, both cloud operators and hackers have ready access to data stored in the cloud. To minimize risks of data interference, data and executable codes must be properly protected through data encryption. However, it is still very difficult to encrypt executable codes since after encryption, executable codes can no longer be read and executed. New techniques with high levels of efficiency are needed to address this problem in the future.

Security information and event management

In recent times, security control devices are becoming more and more sophisticated. They include firewall, proxy, reverse proxy, packet filtering, IDS, IPS, WAF (web application firewall), SEM, SIM, and last but not least, SIEM. SIEM (security information and event management) analyzes the logs from other devices using both template matching patterns and artificial intelligence techniques. If the targeted network is not well protected in general and has malicious traffic floating around, it is very easy to discover hacking traffic and transactions inside the targeted network systems. However, if the targeted network is well protected, it is very difficult to discover anomalies. Although SIEM can only discover malicious traffic if at least one example is present, it is still the best frontline defense for any network as there is no prior assumption of what the traffic format should be.

Social network risks

Online social networks operating in Hong Kong include Facebook, Twitter, WhatsApp, and WeChat. While many financial institutions conduct transactions over WeChat, in Hong Kong, banks refrain from communicating using these services. In practice, many account officers communicate with bank customers for speed and efficiency using such networks, sending photos and recorded voice messages, although at present all discussions, photos, and voice messages can never be official. In the future, we need to explore the use of social network applications with proper data encryption, logging for banking purposes, and reliable social network operators to enhance the security of social media in banking.

Many big data social behavior research projects use data mining techniques, although officially, only social behavior can be identified through data mining research. Unofficially, individual behavior can be examined using the same techniques. As a result, governments and hackers are able to examine the behavior of individuals by studying social networks. Attempts to separate personal from business information in social media tend to be impossible.

Stored-value cards

In Hong Kong, the Octopus card is now used extensively for small value transactions and also for property access authorization and logging. However, owing to the many uses to which Octopus cards are put, every citizen now carries with them multiple cards in their wallets. In most cases, one Octopus card is used for small currency transactions and all the other cards are used for property access authorizations. Although using Octopus cards for small currency transactions is safe, risks can arise if cards are used for multiple other transactions as there is no way of knowing how much currency within the card has been deducted.

Electronic waste disposal risks

Disposal of electronic equipment in financial institutions raises a number of security risks following the appearance of solid state hard disks that can hold up to 0.1 TB of data. Normally, magnetic hard disks can be demagnetized before disposal, but solid state hard disk data can only be erased by direct destruction of the chip itself, which creates serious environmental problems. Efforts are needed by industry to develop effective mechanisms for erasing data stored in chips. Until this has been achieved, financial institutions should avoid using such devices.

Operational procedure risk

Accountability audits should become mandatory in financial institutions as many errors, problems, hacking events, virus attacks, and spamware can be traced back to problems identified in accountability audits. For example, if a desktop PC were attacked repeatedly by a virus, an accountability audit could trace back all possible virus transmission paths in order to determine the source of the virus.

IT document risks

A final area in which security control problems can arise concerns documentation used for design and development of systems. Functional specification is mainly required for predevelopment agreements and for the establishment of design specifications. Design specifications are required for actual design and development as well as for system integration test plans. User guides are mainly needed for users to operate final applications, while operational guides are needed for any extra work that requires some operators to handle problems during end-of-day periods. Guides are also needed for fresh system installation, system migration, data migration, shutdown, startup, and recovery processes. Deployment plans are needed for final production deployment and must be tested in testing environments with simulated real data. Security control specifications must also be developed for both operation and design work. Design specifications, operation guides, and security control specifications will also be used for future enhancement. Providing insufficient documentation will make activities far more difficult while over-documentation will create problems for future enhancement of design and development. During accountability audits, documentation will pinpoint the right direction without wasting valuable time and human resources.

Conclusions

As is apparent from the above discussion, the operation of financial institutions is continuing to become more complex, difficult, and costly. At each point in the operation of a financial institution, security risks arise and need to be managed. It is important to improve the security controls within institutions, despite the fact that some are complex to implement and expensive to deploy. Although security controls can never be perfect, it remains of critical importance for security measures to be continually reviewed and improved, particularly as new systems and procedures are developed. Insufficient budget can never be used as the reason for not taking action.

13
Complicity in Cyberspace: Applying Doctrines of Accessorial Liability to Online Groups

Gregor Urbas

Introduction

Telecommunications technologies have changed not only the ways in which crimes are committed but also the ways in which offenders interact in committing them. The truism of "action at a distance" exemplified by online crime, whereby offenders and victims need not be located in the same place or even the same country, also holds for relationships between co-offenders. Online criminal groups can operate effectively without their members ever meeting in person, being able to recognize each other by sight, or knowing each other's real names.

Three examples illustrate the diversity of such online cooperation:

Example 1: Internet-based copyright piracy groups such as Drink Or Die have involved dozens of globally dispersed members, known only by their screen names, acting together to break copyright protections on business and games software and distribute unauthorized modified versions through known "warez" websites. Their structure is usually highly organized with one or two leaders coordinating others with specialist skills and tasks within the group. Though motivated more by "bragging rights" than financial gain, they knowingly violate the intellectual property laws of numerous countries in their efforts (Urbas 2007, p. 210).

Example 2: Online child exploitation rings are often similarly globally dispersed and organized but increasingly secretive and sophisticated in their use of anti-detection measures such as encryption, anonymizers, and virtual private networks. Child pornography rings such as the W0nderland Club imposed membership requirements of thousands of child exploitation images in order to join, and such groups may use members' ongoing access to fresh images both to assign "rank" with the group and to protect against infiltration by undercover law enforcement officers (Krone 2004).

Example 3: Online "hacktivist" collectives such as Anonymous have large but ephemeral support bases, not all of whose members necessarily appreciate the illegality of some of the group's activities but who use their aggregated strength against identified targets through hacking and distributed denial of service (DDoS) attacks. For example, Anonymous is understood to have been responsible for politically motivated attacks against Australian government websites as part of "Operation Titstorm" in protest against a planned national Internet filter regime (Hardy 2011).

In all of these examples, and many others that could be added (Choo, Smith and McCusker 2007, p. 77), there is a clear force multiplier involved in the online collaboration of groups of individuals. Some groups exhibit structure, organization, and stability over time, while others are more ephemeral and opportunistic associations. Where "command and control" exists, online criminal groups may approximate more traditional organized crime entities (McCusker 2006), and their members may be prosecuted using "criminal enterprise" approaches where appropriate legislation exists (Gilad 2012). In others, the manner of interaction may be less organized but nonetheless amenable to prosecution through the application of legal doctrines of complicity. These are considered below, using the Australian Commonwealth's Criminal Code, the *Criminal Code Act 1995* (Cth), to illustrate the issues.

Legal doctrines of complicity

The general idea behind criminal complicity is that "a person who promotes or assists the commission of a crime is just as blameworthy as the person who actually commits the crime" (Bronitt and McSherry 2010, p. 381). However, the various legal doctrines that have been developed to extend liability to such "accessories" are notoriously complicated, and any sense of a general approach is quickly overtaken by theoretical peculiarities and historical anomalies. For example, the historical distinction between felonies and misdemeanors (now abolished in all Australian jurisdictions) pervaded the way in which accessorial liability was attributed. For more serious offenses (felonies), a person who assisted or encouraged a crime at its commission was a "principal in the second degree," while one who assisted or encouraged beforehand was an "accessory before the fact," and one who helped the offender to evade detection was an "accessory after the fact." However, for more minor offenses (misdemeanors), such fine distinctions in participation were largely ignored.

Aiding, abetting, counseling, or procuring

This compound phrase is usually understood to include various forms of assistance or encouragement given by one person in relation to the crime committed by another. Liability of the accessory is therefore derivative on liability of the principal offender. Traditionally, offenders who were present at the time of another's crime are said to have "aided" its commission if they rendered assistance and "abetted" if they offered encouragement. However, it has been held that all of these words are "instances of one general idea, that the person charged as a principal in the second degree is in some way linked in purpose with the person actually committing the crime, and is by his words or conduct doing something to bring about, or rendering more likely, such commission" (*R v Russell* [1933] VLR 59 per Cussen ACJ at 67, cited by Odgers 2010, p. 163). By contrast, persons not present are said to have "counseled" if they urged or advised the commission of the crime, and "procured" it if they used means such as payment to bring it about.

Some caution, however, needs to be exercised assuming physical presence at the commission of a crime to be a necessary requirement for aiding and abetting, for several reasons. First, despite a series of cases dealing with whether "mere presence" may be sufficient to aid and abet (such as in the prize-fighting case of *R v Coney* [1882] 8 QBD and the concert case of *Wilcox v Jeffery* [1951] 1 All ER 464 cited by Bronitt and McSherry 2010, p. 389), there is no clear authority for physical presence as a necessary condition for accessorial liability.

Rather, the best that can be said is that physically present accomplices are typical of the cases in which aiding and abetting has been charged and successfully prosecuted, while counseling or procuring have been used in relation to accomplices involved before a crime occurs but not present at its commission. Where such persons are actively involved in the crime at the time of its commission but not physically present, for example directing other offenders by means of telecommunications, there would appear to be no reason in principle why they should not be seen as aiding and abetting the offense. This was recognized in the case of *R v Choi (Pong Su No.21* [2005] VSC 96 [21 July 2005] per Kellam J at 59–63), adopting the concept of "constructive presence" as articulated in the English text *Russell on Crime* (notes omitted):

> ... it appears to me that for a person to be found guilty of aiding and abetting the commission of an offence by a principal, requires a link in purpose between that person and a principal. In addition to the link in purpose, the aider and abettor must by words or conduct do something to bring about the commission of the offence. In my view that does not require actual presence.

> ... even if, contrary to my conclusion, presence is required to establish that a person aided and abetted the commission of an offence under the Code, such presence need only be constructive.

In relation to this issue, *Russell on Crime* states:

> It used commonly to be stated that a principal in the second degree must be "present" at the fact: but "present" was always an elastic term and covered the case of a person aiding and abetting at the time even by keeping watch and ward at a suitable distance; and even this was extended by the very convenient term "constructive presence" so that now the true principle is that so long as the accomplice is participating by rendering aid, assistance or even mere encouragement to the actual perpetrator at the very time when the latter is effecting criminal purpose, no matter how far away from the spot he may be, then he is certainly aiding and abetting it, and will be principal in the second degree (if a felony) and punishable just as the principal is in the first degree; if it be misdemeanour he is held guilty of the crime itself. But there must be *some* evidence of encouragement.

In this case, the version of aiding or abetting that was under consideration was in statutory form, in s11.2(1) of the *Criminal Code Act 1995* (Cth). This adopts the common law compound phrase but omits any reference to physical presence (as noted by Odgers 2010, p. 165). Kellam J in *R v Choi (Pong Su [No.21])* went on to observe that, if physical presence at the time of the commission of an offense were necessary under s11.2, then subs (4), which allows termination before the commission of the offense, "would make no sense" (at 61). The better view, therefore, is that complicity through aiding, abetting, counseling, or procuring may be established in relation to co-offenders who are not physically present at the commission of a crime but are otherwise actively involved, for example, through the use of telecommunications.

As regards international legal instruments, Australia is now a signatory to the Council of Europe's Convention on Cybercrime (in force for Australia from 1 March 2013), which is to date the most comprehensive and widely adopted such agreement (Urbas 2013). Apart from detailing substantive offenses and procedural measures that signatory states must incorporate into their domestic laws, the Convention also deals with aiding or abetting:

> **Article 11(1):** Each Party shall adopt such legislative and other measures as may be necessary to establish as criminal offences under its domestic law, when committed intentionally, aiding or abetting the commission of any of the offences established in accordance with Articles 2 through 10 of the present Convention with intent that such offence be committed.

As Australia needed to ensure that its substantive and procedural cybercrime laws complied with the Convention's requirements in order for accession to proceed, it can be taken that the aiding and abetting provision in s11.2 of the Criminal Code does indeed apply to cybercrime offenses. The above observations about the interpretation of "aiding and abetting," and the absence of any requirement of physical presence in s11.2, thus apply to online offending. It remains to be seen whether prosecutors and courts will confirm this reasoning in precedential decisions as a more substantial body of cybercrime cases develops.

Acting in concert / joint commission / common purpose

The common law doctrine of acting in concert attributes primary liability to two or more participants in a crime where their joint conduct brings about the commission of the crime according to an agreed plan but where the evidence is unclear about who played which part in its commission (Bronitt and McSherry 2010, p. 415). The usual application of this doctrine is in circumstances where all co-offenders are present, but it has been recognized that it still applies where one is some distance away (*R v Camilleri* [2001] 119 A Crim R 106).

In the *Criminal Code Act 1995* (Cth), the doctrine finds expression as "joint commission" (s11.2A). This section was not included in the original formulation of the Criminal Code but was added later to remedy a perceived gap in the coverage of its complicity provisions. It depends on there being an "agreement to commit an offense" between two or more persons leading to either commission of the agreed offense by one or more of the parties to the agreement or another offense that is committed by one or more of the parties where one or more of the parties were reckless as to its commission (subs 2 and 3). Importantly, liability imposed through joint commission is principal rather than derivative liability, and it is no bar to conviction of a party to a criminal agreement that any other party was not prosecuted or was not convicted or that "the person was not present when any of the conduct constituting the physical elements of the joint offence was engaged in" (subs 7). This makes it abundantly clear that, unlike the common law doctrine of common purpose, the s11.2A version does not depend on physical presence or even proximity. It is apparent, therefore, that the provision could apply where members of an online group are geographically dispersed but acting together in the commission of an offense.

It should be noted in passing that s11.2 (discussed earlier in relation to aiding and abetting) also encompasses another doctrine of accessorial liability, that of "common purpose" or "joint criminal enterprise." This imposes liability on the participants in an agreed criminal endeavor to any act committed by any of them which falls within the scope of that agreement or understanding. Extended common purpose / joint criminal enterprise takes this further, encompassing not only those acts agreed upon but also those foreseen as possibly being committed in the execution of the criminal plan (Bronitt and McSherry 2010, p. 430).

Innocent agency / commission by proxy

This doctrine allows liability to be imposed on a person who has the requisite fault elements for a crime but does not perform its physical elements, instead duping or manipulating another into doing so. The proxy may be an individual or a corporation (*White v Ridley* [1978] 140 CLR 342, cited by Bronitt and McSherry 2010, p. 412). In the *Criminal Code Act 1995* (Cth), the doctrine finds expression as "commission by proxy" (s11.3).

This doctrine may be applicable to online offending where another person, unwittingly or otherwise, is manipulated into doing an act that results in or facilitates the commission of an offense. A scenario where a "money mule" is used to launder proceeds of crime may provide an example, as in *R v Columbus* (2007) QCA 396, though in this case s474.14 of the Code was used rather than s11.3. It has been argued that this broad offense could be applied to online actors such as WikiLeaks engaged in facilitating the publication of classified material online (Anton and Urbas 2011). Neither the innocent agency/commission by proxy form of liability nor the statutory offense of facilitation of a serious offense through technological means appears limited by any requirement of physical presence. These provisions would therefore apply well to online offending by members of dispersed yet cooperating offenders.

Incitement and conspiracy

Although classified as "inchoate offenses" rather than doctrines of complicity, both incitement and conspiracy involve more than one actor in the commission of offenses. Incitement is the urging of another person to commit a crime, while conspiracy is the agreement between two or more persons to commit a crime. These are found in s11.4 and s11.5 of the *Criminal Code Act 1995* (Cth), respectively. Although incitement is rarely charged, conspiracy is often used by prosecutors in relation to group offending, including in at least one recent cybercrime case (*Larkin v The Queen* [2012] WASCA 238 [23 November 2012]).

In the United States, prosecutions of Drink Or Die members, discussed earlier on, both copyright infringement and conspiracy were charged. In extradition proceedings of Hew Raymond Griffiths ("Bandido"), Australian courts were required to decide whether the acts of the online conspirators, had they occurred in New South Wales, would have satisfied the "double criminality" test. In holding that this test could be applied, the Full Court of the Federal Court (in *Griffiths v United States of America* [2005] FCAFC 34 [10 March 2005]) observed (at 97):

> Whatever may have been the particular place of origin of the conspiracy insofar as it concerned Mr Griffiths, the conduct constituting the offence, given its continuing character, can properly be said to have occurred in the United States and this includes Mr Griffiths's own conduct notwithstanding his actual physical presence in New South Wales.

This importantly provides some judicial authority for the proposition that a person's online activities, particularly when carried out as part of a conspiracy between globally dispersed members of a group, are not physically limited by the location of that person. In other words, the Internet allows "action at a distance," and this extends to complicity in others' crimes (Urbas 2007; Urbas 2012).

In company

A final aspect of complicity to be considered is the aggravating factor of commission of an offense "in company." This provides for increased penalties where an offense is committed in the company of other persons, apart from the victim. For example, s61J of the *Crimes Act 1900* (NSW) is the offense of sexual assault in company, while s61JA is aggravated sexual assault in company, the latter punishable by life imprisonment. The theoretical possibility of using such offenses in the online environment has been noted (Urbas and Choo 2008, p. 42).

This suggestion, however, depends on applying the doctrine of "constructive presence" to the "in company" concept. There are no cases in which this has occurred, but remarks of the New South Wales Court of Criminal Appeal in *R v Button; R v Griffen* ([2002] NSWCCA 159 at 121–125) provide some foundation for the suggestion:

> The physical presence of another is, therefore, required for the crime to be committed in company. However, two questions remain. First, what is meant by physical presence? Secondly, in respect of which aspect of the crime identified by s61J is physical presence required?
>
> ... For the purposes of this appeal, I believe it is really only necessary to answer the first question, that is: what is meant by physical presence, and whether, in the context of this case, the separation of 50 metres was capable of satisfying that requirement?
>
> Physical presence is an elastic concept. The concept is best explained by example. Assume the robbery of a large warehouse, with a number of persons involved. The robbers scatter in search of valuables. They may, at any point, be separated by 50 meters and yet, throughout, will remain "in company." Assume that one robber demanded the bank card and PIN number of the owner. He then separated from his companions, went to the bank one kilometer away, and used the card to withdraw cash before returning to the warehouse. Is the theft of that money part of the crime committed in company? The bank card and the PIN number were, no doubt, obtained with the aid of the coercion exerted by the group.
>
> Take another illustration, closer to the facts in this appeal. Assume a sexual assault in a large house, involving a number of individuals. If, for reasons of privacy, the victim were taken to an adjacent bedroom, and the door closed, the offense would plainly still be one committed in company. And the result, I suggest, would be no different if the bedroom were upstairs, so that some distance separated the offender at the time of penetration and other members of the group.

However, there must be limits. The point must be reached where the separation between the offender and the group is such that the offense can no longer be characterized as being in the presence of a group. How are those limits determined? I believe the learned trial judge accurately identified the test. The test is the coercive effect of the group. There must be such proximity as would enable the inference that the coercive effect of the group operated either to embolden or reassure the offender in committing the crime or to intimidate the victim into submission.

It is arguable that, even though the court in *R v Button; R v Griffen* was dealing with multiple offenders in reasonable physical proximity but still somewhat separated during their criminal conduct, the touchstone of "coercive effect of the group" identified above can be applied to online groups interacting in real time, such as in the live streaming of child sexual abuse on the Internet. The violation of the victim is exacerbated by the group assault in a way that arguably merits recognition through more serious offenses than those relating only to the dissemination of child pornography images.

More recently, Australian law enforcement authorities and courts have begun dealing with "webcam child sex tourism" as a new form of online child exploitation (Terre Des Hommes 2013). This involves an offender in one country paying for and dictating specific sexual acts to be performed in another country, involving children and possibly others, including rape by one or more adults. Such offending can be prosecuted within Australia by charging local offenders with Commonwealth offenses such as using a carriage service for child pornography material, using a carriage service for sexual activity with a child, or using a carriage service to groom a child for sexual activity (Urbas 2008). However, the doctrines of complicity as reflected in the provisions of the Commonwealth Criminal Code, discussed above, might also be applied subject to resolution of jurisdictional issues arising where one offender is in Australia but both the victim and co-offenders are located outside Australia.

Jurisdictional issues

In general, criminal jurisdiction requires a geographical or other "nexus" between the legal system and the crime or offender to exist. Where one country asserts jurisdiction over persons or events outside its borders, this is regarded as "extra-territorial" in its reach. The Commonwealth

Criminal Code has differing levels of jurisdiction for offenses based on geographical nexus.

Standard Geographical Jurisdiction (s14.1) – conduct by a person does not constitute an offense under Commonwealth law unless the following conditions are met:

- the conduct constituting the alleged offense occurs wholly or partly in Australia or on board an Australian aircraft or ship; or
- the conduct occurs outside Australia but a result of the conduct occurs wholly or partly in Australia or on board an Australian aircraft or ship; or
- the offense is ancillary to conduct which occurs, or is intended to occur, wholly or partly in Australia or on board an Australian aircraft or ship.

Extended geographical jurisdiction – category A (s15.1) – conduct by a person does not constitute an offense unless standard geographical jurisdiction criteria applies, or the conduct occurs outside Australia but, at the time of the alleged offense, the person is an Australian citizen or a body corporate incorporated under Australian law.

Extended geographical jurisdiction – category B and category C (s15.2 and s15.3) – similar to category A, but with different conditions relating to defenses.

Extended geographical jurisdiction – category D (s15.4) – conduct may constitute an offense under Commonwealth law whether or not the conduct, the person, or the effects have any connection to Australia (thus amounting to "universal jurisdiction").

The Commonwealth Criminal Code applies these different levels of geographical jurisdiction to different types of offense. For example, federal cybercrime offenses generally attract Extended geographical jurisdiction – category A, while terrorism offenses attract Extended geographical jurisdiction – category D (Urbas and Grabosky 2005; Urbas 2012). In relation to complicity occurring across jurisdictional boundaries, whether the principal offense occurring outside Australia is one that constitutes an offense under Commonwealth law, thus allowing accessorial liability to apply, depends on the specified jurisdictional level. For example, child exploitation offenses may include conduct occurring wholly outside Australia but involving a result that occurs in Australia, such as a transmission of images to a computer in this country or to an Australian citizen or resident. Such conduct will constitute an offense under the Commonwealth Criminal Code, and therefore aiding, abetting, counseling, or procuring this conduct, for example, will also be offenses under the Code.

Conclusion

Information technology facilitates both the commission of cybercrimes and the cooperation of offenders in committing them. This requires a reconsideration of doctrines of complicity as they apply in cyberspace. While some modes of accessorial or group liability (such as incitement and conspiracy) translate easily to the online environment, others (such as aiding and abetting or acting in company) are more challenging as they traditionally require the presence of co-offenders at the commission of a crime. However, the "coercive effect of the group" in online involvement is what is crucial, rather than physical presence, and thus a form of "virtual" or "constructive presence" should suffice for criminal liability to attach.

Bibliography

Anton, D. and Urbas, G. (2011) "Why Julian Assange May Have a Case to Answer in Australia," *Legal Theory Blog,* http://papers.ssrn.com/sol3/papers.cfm?abstract_id=1733666## (accessed 11 February 2015).

Bronitt, S. and McSherry, B. (2010) *Principles of Criminal Law* 3rd edn, Sydney: Thomson Reuters.

Choo, K-K. R., Smith, R. G. and McCusker, R. (2007) "Future Directions in Technology-Enabled Crime 2007–09," *Research in Public Policy Series* 78, Canberra, Australia: Australian Institute of Criminology.

Gilad, M. (2012) "Virtual or Reality: Prosecutorial Practices in Cyber Child Pornography Ring Cases," *Richmond Journal of Law and Technology* 18: 1–66.

Hardy, K. (2010) "Operation Titstorm: Hacktivism or Cyber-Terrorism?," *University of New South Wales Law Journal* 33(2): 474–502.

Krone, T. (2004) "A Typology of Online Child Pornography Offending," *Trends & Issues in Crime and Criminal Justice* no. 279, Canberra, Australia: Australian Institute of Criminology.

McCusker, R. (2006) "Transnational Organised Cyber Crime: Distinguishing Threat From Reality," *Crime, Law and Social Change* 46: 257–73.

Odgers, S. (2010) *Principles of Federal Criminal Law* 2nd edn, Sydney: Thomson Reuters.

Terre des Hommes (2013) "Webcam Child Sex Tourism," http://www.terredeshommes.org/webcam-child-sex-tourism-2/ (accessed 11 February 2015).

Urbas, G. (2007) "Cross-National Investigation and Prosecution of Intellectual Property Crimes: The Example of 'Operation Buccaneer'," *Crime, Law and Social Change* 46: 207–21.

Urbas, G. (2008) "Look Who's Stalking: Cyberstalking, Online Vilification and Child Grooming Offences in Australian Legislation," *Internet Law Bulletin* 10(6): 62–9.

Urbas, G. (2012) "Copyright, Crime and Computers: New Legislative Frameworks for Intellectual Property Rights Enforcement," *Journal of International Commercial Law and Technology* 7: 11–26.

Urbas, G. (2012) "Cybercrime, Jurisdiction and Extradition: The Extended Reach of Cross-Border Law Enforcement," *Journal of Internet Law* 16(1): 7–17.

Urbas, G. (2013) "Australia's Accession to the Convention on Cybercrime," *Internet Law Bulletin* 16: 2–3.

Urbas, G. and Choo, R. (2008) "Resource Materials on Technology-Enabled Crime," *Technical and Background Paper* no. 28, Canberra, Australia: Australian Institute of Criminology.

Urbas, G. and Grabosky, P. (2006) "Cybercrime and Jurisdiction: An Australian Perspective," in *Cybercrime and Jurisdiction: A Global Survey*, B.-J. Koops and S. Brenner (eds.), IT and Law Series no. 11, The Hague, the Netherlands: TMC Asser Press.

14
Profiling Cybercrime Perpetrators in China and its Policy Countermeasures

Darrell Chan and Dawei Wang

Acknowledgments

The authors would like to thank Peter Grabosky, Roderic Broadhurst, Brigitte Bouhours, Mamoun Alazab, and Steve Chon for their advice and comments on an earlier draft.

Introduction

China has the largest online population of any nation in the world. However, this has been accompanied by serious cybercrime problems. In 2012, police nationwide solved more than 118,000 cases involving networked crimes, with 216,000 suspects being arrested (Wei 2013). A comprehensive analysis (Wei 2013) based on four indices relating to victims, amount of money involved, severity, and influence shows that online fraud, online pornography, and online pyramid schemes are the top three cybercrime types in China. Whether in terms of scale, scope, or the extent of loss, China is one of the countries most impacted by cybercrime. In recent years, much cybercrime research has focused on technologies and victims but less on perpetrators. The absence of evidence concerning the key characteristics of perpetrators, as well as the extent, structure, and activities of cybercrime groups impedes the development of sound countermeasures.

This chapter seeks to fill a gap in research on cybercrime perpetrators in China. Specifically, it has three objectives. The first is to summarize some of the key features of individual perpetrators. The second is to analyze the structure of cybercrime groups and how such structures influence the processes of group formation and facilitation of criminality. The last objective is to discuss some feasible strategies to prevent and control cybercrime in China based on the characteristics of perpetrators.

Perpetrators' demographic characteristics

Using available official and academic research statistical data, this section profiles the characteristics of "typical" cybercrime perpetrators in terms of their location, demographics, and motivations.

Suzhou city

Recent empirical research in Suzhou city, one of China's most economically developed cities, has found that 120 cases involving 195 offenders were recorded by the prefecture's judicial and procuratorial agencies between 2007 and 2010 (Li et al. 2011). It was found that 91.28 percent of offenders were males, 81.03 percent were between the ages of 18 and 35, and 15.90 percent were aged between 36 and 50 years. Regarding the level of education, 36.6 percent were college educated or above, 35.05 percent had completed junior high school, and 21.13 percent had completed a senior high school education. The top three occupations of offenders were unemployed (40%), employees of enterprises (28.72%), and manual workers (peasants) (13.85%). Over one in five (23.3%) cases involved joint offenses with more than one offender. In most cases, the Internet was used as an instrument or facilitator for the commission of the crime (99.13%). Profit was the most prominent motivation in Suzhou city.

Guangdong province

A study by Zhang (2007) of cybercrime in Guangdong province, the most economically developed province in China, showed that property-related crimes involving the Internet were the most common cybercrime type between 2004 and 2006. The average amount involved in each case was 65,900 RMB Yuan (US$10,740) over three years. Offenders were overwhelmingly male (94.8%), and most were aged between 18 and 25 years (54.4%). One-third (32.8%) were between 26 and 59 years, and those under the age of 18 accounted for 12.5 percent. As to educational level, nearly half of the offenders (49.5%) had a senior high school level of education or below; 27.9 percent had reached junior college/bachelor degree; and 22.6 percent were college graduates or above. Data demonstrated that joint offending was increasing, from only 19.4 percent in

2003 to 27.8 percent in 2004, 27.5 percent in 2005, and 32.6 percent in 2006. The number of joint offenses in 2006 was more than the sum of those recorded in all of the preceding five years.

Gender and age

In terms of gender and age, offenders in Suzhou city and Guangdong province were overwhelmingly male and younger in age (18–35 years). Another data set from the Ministry of Public Security reported that of known cyber offenders in 2005, 45 percent were between 18 and 25 years of age (Zhang and Zhang 2005). In addition, recent data from Hubei province for 2011–12 (Zeng and Wang 2012) showed that 90 percent of known cybercriminals were under 30 years of age. Compared with international research data, Li's (2008) research showed that male offenders constituted the absolute majority of cybercriminals, who were primarily between the ages of 17 and 45 years. This can be explained by the fact that people in this age group have been profoundly affected by the development of modern information technology, and surfing the Internet has become a lifestyle for them. The number of offenders in this age group is proportional to the number of Internet users in this age group. We can predict that along with the development of computer science and the growth of Internet popularity, the age of offenders is likely to increase.

Professional qualities and occupation

Perhaps surprisingly, levels of educational attainment and technical skills among cybercriminals have been found to be lower than expected, while levels of unemployment among cybercriminals tend to be high. It is commonly assumed that cybercrime offenders are people who have a high level of education and technical skills. But statistics suggest the educational levels of cybercrime offenders are lower than expected. In Suzhou city and Guangdong province, about half of the perpetrators had a senior high school education or below. Similar trends can be found in international studies. For example, Marcum, Higgins, and Tewksbury (2012) found that the average educational level of cyber offenders was a high school diploma. Often, those with lower levels of education and professional skills can still invade information systems by means of hacker toolkits and botnets that are readily available online. Another significant characteristic of cybercriminals in China is unemployment. For example, in Suzhou province, 40 percent of offenders were unemployed.

These findings can be explained in two ways. The threshold of cybercrime sophistication has been significantly reduced by the wide-scale use of toolkits and botnets, which allow offenders without professional skills to easily access computer systems illegally. This dramatically lowers the skill levels required for offending. With the formation of cybercrime "industrial chains," some skilled activities and tools can be rented or bought on the Internet, making high levels of technical proficiency unnecessary. On the other hand, in many cybercrime groups, only a small number of skilled professionals are in charge of technologically sophisticated offending. As expected, with the growth in popularity of the Internet and the development of information technology, especially the rapid penetration of mobile Internet, this threshold will continue to decline, and people from all skill levels will be able to be involved.

Motivations

Profit is clearly the most common driving force behind the commission of cybercrime. With the rapid growth of e-commerce, the thrill-seeking computer-savvy hackers of earlier times with their quasi-ideological culture have been supplanted by the benefit-chasing cybercriminals of today. In 2012, 46.1 percent of nationwide registered cybercrime cases concerned online fraud (Gu 2013). In addition, two reports from Shenzhen city in 2011 and Luoyang city from 2006–09 suggested that online fraud ranks as the most common of all cybercrime types – 57 percent and 69.9 percent respectively (Dong, Ma and Yang 2012; Zhang 2010).

This can be explained in terms of the cost-effectiveness of cybercriminal activities. As we know, rational criminals tend to take into account costs and benefits in determining whether and how to offend. The Internet provides opportunities to decrease the cost and maximize the benefits of crime. In relation to costs, victims' lack of awareness and technical protection enables offenders to easily invade and steal their information. Moreover, to any rational offender, "... [L]egal punishment is the largest cost that criminals will take into consideration, not only the severity of any punishment, but also the possibility of being arrested, prosecuted and convicted" (Chang 2012, p. 86). The lower risk of being identified and sanctioned significantly reduces the cost for offenders. Research from Guangzhou Intermediate People's Court showed that from 2008 to 2011, the number of sentenced online fraud cases was less than 1 percent of the reported number (Dong, Ma and Yang 2012). Accordingly, the costs associated with cybercrime appear to be low.

In terms of benefits, with the advent of cloud computing and booming e-commerce, more and more opportunities arise for cybercriminals to secure a financial profit through offending merely by the click of a mouse. The risks are much less than in the case of robbery or theft in the terrestrial world, while the benefits are much greater.

Analysis of organized cybercrime groups

Article 2 of the *United Nations Convention against Transnational Organized Crime* defines an organized criminal group as:

> ... [A] structured group of three or more persons, existing for a period of time and acting in concert with the aim of committing one or more serious crimes or offences established in accordance with this convention, in order to obtain, directly or indirectly, a financial or other material benefit (United Nations 2004).

We can find all these characteristics in organized cybercrime groups. No doubt some cybercrime may involve conventional organized crime groups that have adapted to cyberspace. They most likely reflect patterns in the conventional world while deeply influenced by digital technology as well. One of the most noteworthy characteristics of the Internet is the way it can empower individuals and, of course, it can empower organizations as well. The exponential growth in connectivity thus accelerates and enhances the capacity of cybercrime groups. As Grabosky (2007, p. 156) said, "cybercrime in general and organized cybercrime in particular are following three basic trends: sophistication, commercialisation, and integration."

There is a lack of systematic research and statistics about the nature of cybercriminal organizations active in China, but one apparent trend is that "conventional" organized crime groups have become increasingly involved in cybercrime.

Social networking services (SNS)

The emergence of SNS in China such as Wenxin, Weibo, and QQ has enhanced connectivity and provided new opportunities for offenders. On one hand, group members can coordinate by SNS anytime, anywhere, more conveniently, and less visibly. On the other hand, it has enabled groups not only to commit crimes more deceptively but also to extend to the victims' family members and friends. It has also accelerated the growth of cybercrime in the mobile Internet. While 87.3 percent of Internet users in China have installed security protection software, only 41.1 percent of smartphone users use such software (China Internet Network Information Centre 2012), which places them at high risk.

Cybercrime group structure

Compared with traditional organized groups who have a hierarchical top-down command-and-control structure, the overall structure of cybercrime groups tends to involve a loose network or even a peer-to-peer decentralized structure. With respect to the composition of members, the connections between them are looser and more fluid. To some extent, individual members can join or exit freely. Another noticeable point is that members come from more diverse backgrounds and may be geographically dispersed. Unlike traditional groups based on interpersonal relationships, members of cybercrime groups may never meet each other during the life of the group's activities. This flexible, loose structure has at least two advantages. First, it promotes the efficiency and the benefits to be derived from crime while obviously reducing the cost of the crime to the offenders. Moreover, it makes the whole group less visible. Even if some members have been arrested, there will be less impact on the whole group, and it can undertake "self-healing" quickly, making it difficult for law enforcement to identify, obtain evidence, and eliminate groups thoroughly.

Specialization

Driven by the lure of tremendous profit, an industrial chain is emerging that consists of making hacker toolkits, undertaking illegal network attacks, securing illegal access to and theft of information, selling information, carrying out fraud, and laundering the proceeds of cybercrime. Each link in this chain has specialized offenders, some even with professional training. As shown in finalized prosecution cases, offenders have a clear division of labor, coordinate activities with each other, and collectively complete the entire process of crime. The emergence of internal differentiation and specialization improves the efficiency and extends the scope and capacity of the criminal group. It also increases the difficulty of tracing and combating them.

Cross-border activities

In recent times, more and more cybercrime groups commit their crimes across jurisdictional borders. As Brenner (2006, p. 190) argued: "cyberspace and computer technology make geography irrelevant." From 2004 to 2010, Chinese police have assisted 41 countries in the investigation of 721 cybercrime cases (Statement by the Chinese Delegation 2011). Statistical data show a rapid rise in cross-border cybercrimes, mainly focusing on online fraud, online gambling, cyber pornography, and hacker attacks. Crime groups use the transnational characteristic of the Internet to evade legal sanction. Due to the lack of international cooperation and assistance, as well as the complex, time-consuming process that seeking mutual legal assistance can entail, groups can commit large-scale, persistent, and very lucrative crimes with little chance of interdiction.

The transnational characteristic of cybercrime makes it difficult to identify the whole crime chain and secure sufficient evidence to undertake prosecutions. It also ensures that cybercrime groups have significantly more impact than traditional organized crime groups. It is worth noting the problem of so-called "jurisdiction shopping" in which groups take advantage of differences in national legislation to commit crimes in countries with the least effective and punitive judicial systems, thus enabling many to avoid prosecution and punishment.

The involvement of legitimate enterprises in cybercrime

In the past, the alleged perpetrators of cybercrime were mostly natural persons. However, the emergence of e-commerce has seen some legitimate enterprises become involved in the manufacture and sale of counterfeit items over the Internet and other intellectual property crimes. According to the Dongguan City People's Procuratorate, in 2012, 22 percent of defendants prosecuted in counterfeiting cases conducted their activities under the cover of legitimate corporations or factories (Zhou 2013). These legitimate enterprises are more organized and more specialized than their illicit counterparts, and their lawful appearance makes them much harder to detect. Local protectionism appears to be an important factor here. Some local governments pursue economic growth excessively, ignoring the obligations of intellectual property rights and tolerating counterfeiting without intervention. According to Ma (2013), government officials who tolerate counterfeit activities should be held criminally responsible.

Policy countermeasures: A comprehensive prevention and control framework

It is a common view that control of cybercrime requires joint efforts and a comprehensive approach. As mentioned in a resolution adopted by the UN General Assembly on the creation of a global culture of cybersecurity (United Nations 2003): "effective cybersecurity is not merely a matter of government or law enforcement practices, but must be addressed through prevention and supported throughout society."

Based on routine activity theory (Cohen and Felson 1979), and employing the primary/secondary/tertiary crime prevention model (Brantingham and Faust 1976) and the two-dimensional typology of the crime prevention model (Van Dijk and Waard 1991), this chapter introduces a framework to prevent and control cybercrime in China, the elements of which are presented in Table 14.1.

Table 14.1 A comprehensive countermeasures framework of cybercrime prevention and control

Target Groups	Developmental Stage of Cybercrime Prevention		
	Primary (General Public)	**Secondary (Risk / Core Groups / Situations)**	**Tertiary (In / After Cybercrime)**
Offenders	1 Various Forms of Education	2 Identification and Guidance	3 Introduce Qualifications Punishment
Situations	4 Mass Media, Domestic Legislation, Internet Industrial Safety Protection	5 Wide Range of Cooperation, More Flexible International Cooperation and Assistance Mechanisms	6 Emergency Response & Cooperation Mechanism, Technical Support, Transnational Law Enforcement Cooperation
Victims	7 Self-Awareness, Technical Measures	8 Personal Information Protection, Online Trading Protection, Guidance for Youth	9 Cooperation with Law Enforcement, Legal, Technical, Psychological Assistance

From Table 14.1, inspired by the two-dimensional typology of crime prevention model, two dimensions are apparent. The first dimension is based on the theory of primary/secondary/tertiary prevention but merges risk/core groups and situations into secondary prevention, and countermeasures during or after crimes have been committed are situated in the tertiary stage. The second dimension is inspired by routine activity theory. It divides prevention countermeasures into offender-oriented, situation-oriented, and victim-oriented categories. The combination of these two dimensions leads to a subdivision of cybercrime prevention and control into nine different types, as shown in Table 14.1.

According to this model, offender-oriented prevention is targeted at the public as potential offenders and seeks to identify risk groups and give them guidance. Aimed at actual offenders, it entails criminal punishment as well as administrative and regulatory remedies.

Situational-oriented prevention focuses on policy and legislative countermeasures to improve Internet security. Based on the actual situation, a wide range of cooperation and emergency response mechanisms are emphasized to respond to cybercrime more efficiently.

One of the important aspects of cybercrime prevention is victim-oriented prevention, targeted at the public at large as potential victims. The goals are to improve self-awareness of prevention, to give guidance to groups and industries at risk, and most importantly, to call attention to the necessity of protecting personal information.

A brief discussion of the framework

Offender-oriented prevention

Primary offender-oriented prevention (category 1)

In the primary stage, members of the public are regarded as potential offenders. Countermeasures are aimed at strengthening inhibition to commit cybercrimes by potential offenders. Such programs include various kinds of education: by school, community, mass media, and family. The core purpose of this education is to cultivate the correct behavior standards on the Internet and enhance knowledge of law.

Secondary offender-oriented prevention (category 2)

This stage is based on early identification and specific guidance to risk groups. From the empirical data presented above, youth and the unemployed are primary risk groups as well as individuals holding some high-risk positions in financial and online trading industries. Particular guidance such as job training, legal education, improving self-esteem, and self-discipline need to be provided to these individuals. In addition, special management including identification, recognition, parent/school/community guardian, and community correction measures should be taken to prevent potential crimes.

Tertiary offender-oriented prevention (category 3)

By the seventh amendment to the Criminal Law in 2009, China has further improved the substantive law on cybercrime. However, it is still unfortunate that punishment based on limiting access to digital technology has not been introduced. So a more flexible punishment regime needs to be adopted, including not only deprivation of liberty but also deprivation of property and qualifications (i.e., Internet access) as forms of punishment. It may be necessary to consider depriving access to computer-related activities or prohibiting engagement in a related vocation in the long or short term in appropriate cases for some cybercriminals.

Situation-oriented prevention

The approaches discussed in this sector seek to make cybercrime more difficult or costly for potential offenders. Countermeasures listed here require joint efforts by law enforcement, enterprises, the private sector, and society as a whole. International cooperation plays an important role in both prevention and control.

Primary situation-oriented prevention (category 4)

Situational prevention at this stage emphasizes the external environment. For the public at large, both potential offenders and victims, mass media (especially networked social media) can provide basic knowledge, skills, standards of correct online behavior, and preventive measures directly, at a low cost, and in a form that can be accepted widely. It is also suggested that further improvements be made to substantive and procedural laws, especially concerning the collection and use of electronic evidence and its legal validity. Meanwhile, Internet safety protection must be enhanced, since all the enterprises in the industrial chain have the responsibility for building a safe environment by enhancing comprehensive technological measures for all Internet users.

Secondary situation-oriented prevention (category 5)

Prevention of cybercrime cannot be effective without cooperation among different countries. Since China is not a signatory to the Council of Europe Convention on Cybercrime (also known as The Budapest Convention), we propose more flexible international cooperation and assistance mechanisms at the present stage.

First are bilateral or multilateral policing cooperation. China has established bilateral policing cooperative relationships with more than 30 countries' law enforcement agencies and 49 police liaison officers posted to 27 countries. These resources should be used to strengthen policing cooperation in information exchange and law-enforcement activities. Cybercrime investigation contact mechanisms should be established when conditions are appropriate. Secondly, China must take full advantage of mutual assistance treaties to enhance international cooperation. In recent years, mutual assistance requests on cybercrime have mainly related to online fraud, child sexual exploitation, and infringement of intellectual property rights, and these have increased noticeably. China has offered assistance consistent with relevant treaties. On this basis, further bilateral judicial assistance should be extended and strengthened. Thirdly, there should be improved working relationships with international organizations such as Interpol, SCO (Shanghai Cooperation Organisation), and APEC (Asia-Pacific Economic Cooperation) to share information, conduct law enforcement cooperation, and carry out technical exchange. In addition, it should be stressed that in the long run, China should attend or participate in drafting an international cybercrime convention to prevent and combat cybercrime more efficiently.

Tertiary situation-oriented prevention (category 6)

Prompt action is the key point in this stage. From a law enforcement perspective, enhanced transnational cooperation should be encouraged. Current levels of cooperation in law enforcement or private enterprise are far from sufficient. What we need is cooperation among government agencies, law enforcement, Internet enterprises, Internet service providers, the private sector generally, and academic institutions from all aspects including policy, legislation, enforcement, technology, and morality. So we propose the integration of all resources to establish an interactive emergency response and cooperation mechanism nationwide, and we call for technical support from a variety of professional organizations. Figure 14.1 shows a model of such a mechanism.

This mechanism is based on four principles outlined below: initiated, maintained, and supported by government, a uniform interactive

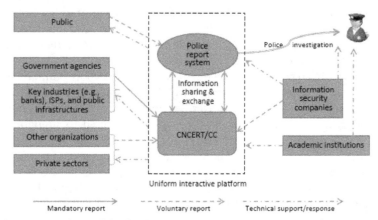

Figure 14.1 A model of cybercrime emergency response and cooperation mechanism

platform which is responsible for receiving information and providing feedback to all stakeholders, combining voluntary and mandatory reporting and appropriate incentive and reward mechanisms.

A uniform interactive platform is the center of the model. All the reports, responses/feedback, and any related information should be made through this platform. Considering the present situation and practical effects, a platform consisting of CNCERT/CC and police report systems is suggested. As the state-level Internet security emergency response center, CNCERT/CC has a comprehensive capability of Internet security incident monitoring, analyzing, and emergency response. We propose that all stakeholders such as government agencies, financial and commercial institutions, information security companies, and private sectors should work together to contribute and report information on security incidents to CNCERT/CC. The public can report directly to police as usual. There would be information sharing and exchange mechanisms between the two systems. For example, all the information on criminal cases could be dealt with by police through the police reporting system, while others are analyzed and responded to by CNCERT/CC through the platform.

In this model, mandatory reporting should be required for government agencies and some key industries such as banks, ISPs, and public infrastructure organizations. Meanwhile, voluntary reporting should be encouraged for other stakeholders. In order to increase the reporting rate, appropriate incentives and awards are needed. Compared with material awards, flexible "non-monetary" incentives, such as information sharing, technical support, and reputation reward are highly recommended.

Arguably, this is a feasible and economical way to solve the problem of police and government agencies' lack of manpower and technical capacity by encouraging academic institutions and information security industries to join the platform. They have the capacity to provide technical support and guidance to victims as well as help law enforcements to investigate cases through the platform. Good relationships and trust are also important in this model. A more equal and harmonious relationship among all stakeholders is advocated to encourage information sharing.

Victim-oriented prevention

Victim-oriented prevention is always the key point in this framework. Potential victims' self-awareness and self-protection are more important than any other technology or legal measures.

Primary victim-oriented prevention (category 7)

By means of the education system and mass media, the public should improve levels of self-protection awareness regarding their Internet activities, and government should put this into the school curriculum so that members of the public are better educated from a young age. In addition, technical measures such as anti-virus software or firewalls should be encouraged.

Secondary victim-oriented prevention (category 8)

In recent years, many cybercrime cases have involved leaking of personal information. In this digital era, our sensitive personal information is widely stored in government, financial institutions, and other commercial organizations' computer systems, and our activities can be easily traced by social networks. On the one hand, the public must be sufficiently aware to protect their own information and take appropriate protective measures. On the other hand, the government has the responsibility to improve legislation on personal information protection and to supervise its performance. In particular, the following measures could be considered.

Strengthening self-protection awareness. Strengthening self-protection awareness is the first and most important rule and should be kept in mind during all online activities. The use of effective technology protection measures such as anti-virus software and firewalls is highly encouraged, especially in online trading. It is also necessary for schools and families to provide guidance for youth to help them reduce the potential risk of being victimized on the Internet.

Improving personal information protection legislation. In China, legislation on personal information or privacy protection tends to be fragmented, and such limited and indirect protection cannot satisfy the needs in the information era. The improvement of legislation should focus on clearly defining the scope of protected personal information, giving classification protection, and most important, the provisions of the law should be as specific as possible to enhance its effectiveness so as to ensure the rights of citizens to protect their information.

Encouraging industrial self-regulation. Internet companies and related industries have a particular and important role to play in relation to Internet security and personal information protection. Various forms of self-regulation should be encouraged and strengthened. It is recommended that government adopt a policy of "less intervention, more self-regulation" by guiding industry associations and by establishing explicit specifications and standards of behavior.

Technology prevention is the last defense of personal information protection. On the one hand, individuals should install anti-virus/spyware software or firewalls and update security patches promptly. On the other hand, the Internet security industry should strengthen its technologies, develop new products, and give targeted guidance to users.

Tertiary victim-oriented prevention (category 9)

The large dark figure of unreported crime is always a prominent problem of cybercrime. Many individuals are unwilling to report cases involving small losses or unsuccessful criminal attempts. For some enterprises or financial companies, the potential harm to business reputation makes them unwilling to report. All these lead to a low reporting rate. To avoid this vicious circle, mutual cooperation between victims and relevant organizations is highly encouraged. First, it is important to encourage victims to cooperate with law enforcement to report cases in time and to provide details as far as possible. Secondly, a comprehensive assistance mechanism should be established to give victims targeted assistance including legal, technical, psychological, and public relations. Thirdly, the privacy of some victims such as banks or companies that handle sensitive personal information should be protected to reduce their concern over risk to "commercial reputation."

Finally, proper incentive mechanisms could be introduced to encourage victims to report. A good example is a first-time-free service provided in Malaysia to both the government and private-sector reporters to help them recover from computer damage. They may charge fees for any subsequent reports of damage (Chang 2012, p. 210).

It must be pointed out that the nine categories in this comprehensive prevention model are not isolated and have no clear time limits. In most circumstances, a well-chosen combination of these categories will be much more effective than isolated measures, and the optimal combination will depend on the particular type of crime threat.

Conclusion

The Internet has seen everyone and every country connected more closely than ever. Unfortunately, it enhances the capacity of criminals as well. It appears that in the foreseeable future, there will be a persistent increase of cybercrime in China. Perpetrators will continue to include people of widely diverse ages and backgrounds, and the threshold of committing crime will become even lower. Empirical research has confirmed a rapidly increasing tendency of cybercrime to be committed by organized groups, and a complete industrial chain of crime has been apparent.

Cybercrime prevention requires participation of the whole society. The two-dimensional comprehensive framework is an attempt to draw from some mature crime prevention models and combine the present situation in China. In this framework, a more flexible international cooperation and assistance mechanism and a national interactive emergency response and cooperation mechanism are highlighted. This chapter also maintains that personal information protection is a key factor in cybercrime prevention and that self-protection awareness is the first and most important rule.

In general, we still need to further both theoretical application and empirical analysis in order to achieve a more in-depth and comprehensive understanding of cybercrime perpetrators. Finally, Internet users should be encouraged to participate more fully and effectively in cybercrime prevention in order to provide for a harmonious online world.

Bibliography

Brenner, S. (2006) "Cybercrime Jurisdiction," *Crime, Law and Social Change* 46(4–5): 189–206.

Brantingham, P. J. and Faust, F. L. (1976) "A Conceptual Model of Crime Prevention," *Crime & Delinquency* 22: 284–96.

Chang, Y. C. (2012) *Cybercrime in the Greater China Region: Regulatory Responses and Crime Prevention Across the Taiwan Strait*, Cheltenham, UK: Edward Elgar Publishing Limited.

China Internet Network Information Centre (2012) *Research Report on Information Safety Status of Chinese Internet Users in 2012*, Beijing: China Internet Network Information Centre.

Cohen, L. E. and Felson, M. (1979) "Social Change and Crime Rate Trends: A Routine Activity Approach," *American Sociological Review* 44: 588–608.

Dong, L., Ma, W. F. and Yang, Y. J. (2012) "Lack of Effective Means in Crackdown Online Shopping Fraud Crime," *Yangcheng Evening News*, 3 September, A8.

Grabosky, P. (2007) "The Internet, Technology, and Organized Crime," *Asian Journal of Criminology* 2(2): 145–61.

Gu, J. (2013) "Online Fraud Accounts for 46.1% of Cyber Criminal Cases in 2012," http://news.xinhuanet.com/fortune/2013-01/23/c_124271076.htm (accessed 16 July 2013).

Li, X. G. (2008) "The Criminal Phenomenon on the Internet: Hallmarks of Criminals and Victims Revisited through Typical Cases Prosecuted," *University of Ottawa Law & Technology Journal* 5(1–2): 125–40.

Li, Z., Jin, C. X., Zhang, F. J. and Yang, M. (2011) "Survey and Analysis on Cybercrime from 2007–2010 in Suzhou City," *Journal of Criminal Science* 10: 120–7.
Ma, X. (2013) "Eliminate Local Protectionism, Consistent Fighting Against Illegal Activities such as Infringing upon Intellectual Rights and Manufacturing and Marketing Fake Commodities," http://www.ipraction.cn/2013/02/16/ARTI1361008468220148.shtml (accessed 10 October 2013).
Macrum, C. D., Higgins, G. E. and Tewksbury, R. (2012) "Incarceration or Community Placement: Examining the Sentences of Cybercriminals," *Criminal Justice Studies* 25(1): 33–40.
Statement by the Chinese Delegation on the International cooperation Issue at the 1st UN Expert Group Meeting on Cybercrime (2011) http://www.chinesemission-vienna.at/chn/hyyfy/t790751.htm (accessed 22 September 2013).
United Nations (2003) *Resolution adopted by the General Assembly 57/239. Creation of a Global Culture of Cybersecurity* (A/RES/57/239), New York: United Nations.
United Nations (2004) *United Nations Convention Against Transitional Organized Crime and the Protocols Thereto,* New York: United Nations, http://www.unodc.org/documents/treaties/UNTOC/Publications/TOC%20Convention/TOCebook-e.pdf (accessed 12 February 2015).
Van Dijk, J. J. M. and De Waard, J. (1991) "A Two-dimensional Typology of Crime Prevention Projects; With a Bibliography," *Criminal Justice Abstracts* 23(3): 483–503.
Wei, Y. Z. (2013) *2012 Annual Report on Cybercrime in China,* Beijing: Chinese People's Public Security University, Policing Reform and Development Research Centre.
Zeng, C. and Wang, X. (2012) "Preliminary Results Were Achieved in Cracking Down Cybercrime by Hubei Police: 2670 Suspects Captured," http://www.chinanews.com/fz/2012/10-23/4270201.shtml (accessed 18 June 2013).
Zhang, J. X. and Zhang, Y. N. (2005) "Sharply Increasing Cybercrime Starts to Merge with Traditional Crime," http://www.china.com.cn/news/txt/2005-11/18/content_6034369.htm (accessed 20 September 2013).
Zhang, W. J. (2010) *Empirical Research on Cybercrime of Metropolis,* master thesis. Beijing: China University of Political Science and Law.
Zhang, X. W. (2007) "An Empirical Research on Property-related Crime Over Internet in Guangdong Province," *Journal of Criminal Science* 4: 95–101.
Zhou, W. L. (2013) "30% of the Cases of Production and Distribution of Counterfeit Sued in Dongguan City People's Procuratorate in 2012 Were Related to Tobacco and Alcohol," *Guangzhou Daily,* 15 March, A5.

Part IV
Privacy and Freedom Online

15
When Privacy Meets Social Networking Sites – with Special Reference to Facebook

Yachi Chiang

Introduction to social networking and Facebook

One of the developing enablers of cybercrime in recent years is the use of social media. Chan, Goggin, and Bruce (2010, p. 593) found the term "social media" to be "a rather broad and vague term for a range of new Internet technologies that work on principles of social networking and other new forms of association." They went on to explain that social networking sites (SNS) "link friends and acquaintances, providing ways not only to create very large groups and networks, but new ways of presenting the self, and communicating with others" (Chan et al. 2010, p. 593).

Boyd and Ellison (2007) have identified a number of characteristics of SNSs including allowing users to establish public or semi-public personal profiles, enabling users to build up specific connections with other users, and permitting users to view and share information posted on SNSs. These three characteristics also reflect the three functions of SNS described by Grimmelmann (2009, p. 4): building up self-identity by allowing users to express themselves through building personal profiles, building up relationships with other users through the use of SNS, and building up the community by permitting users to establish social networks through exchanging information on the SNS.

The history of social networking can be traced back to the 1990s, although many SNSs developed earlier than this (Grimmelmann 2009, p. 5). Among many SNSs, Facebook, which was established in 2004, and although not the earliest pioneer in the field, is so far the biggest SNS with the largest number of users. Along with other attributes,

Facebook has included more and more functions in its platforms over time. Facebook not only permits traditional SNS activities, such as allowing users to establish personal profiles, post, or share personal updates, news linkages, or video clips, it also provides integrated services similar to other applications such as email systems, online chat rooms, and user blogs. It also seeks to maintain its user network by allowing users access to other applications including games and psychological tests.

Based on the design of Facebook's platform, users may disclose their own or others' personal information in a variety of ways.

Establishing a personal profile

There are three main types of personal information that can be present in user profiles on Facebook: basic personal information such as gender, hometown, birthday, and so on; contact information such as email address, instant messenger account, personal website, or office address; and personal information such as users' habits, favorite activities, personality descriptions, and so on.

Posting on the timeline wall

Users can also publish posts on their own timeline walls and allow others to see or share them. The content of posts can include texts, images, linkages, or audio/video files. Users may disclose information concerning their own, or their friends', lives in posts and may also use "tagging" functions in photos by attaching other users' Facebook account names on the individual shown in the photo so that other viewers can easily identify who is in the photo.

User blogs

Users may also disclose personal information about themselves or others' lives by writing blogs.

Psychological test results

By completing online psychological tests on Facebook, users may disclose their own personal information on the timeline wall or allow this information to be provided to application developers based on the default terms and conditions of the tests.

Developments in Facebook's privacy settings

When Facebook was launched in 2004, it was limited to the students of Harvard University in the United States. Gradually, Facebook expanded to other universities in the United States, but it was not until 2006 that Facebook opened itself to anyone who is interested in being a user (Boyd and Hargittai 2010). Despite being the most popular SNS, Facebook is not the only social networking option. In 2006, for example, 80 percent of first-year freshmen in the United States chose to use Facebook, while 54 percent chose to use MySpace (Hargittai 2007). Nonetheless, the most distinctive feature of Facebook is its privacy control options designed for users (Boyd and Ellison 2007).

In the beginning, when Facebook was operating from Harvard University campus, its networking concept was limited to this specific campus and its privacy settings were, by default, set to allow users to see other users' personal profiles if they studied at the same university. Later, when Facebook expanded to corporate networks and district networks, Facebook began to provide more privacy setting options for users. For example, users could permit their disclosed information to be seen by "friends," "friends of friends," or "friends of specific network."

After expanding the idea of user networking, Facebook further introduced other applications developed by the third parties, allowing these applications to share users' personal data through the Facebook platform with the consent of users (Boyd and Hargittai 2010). While Facebook is becoming more and more multifunctional, privacy issues on Facebook have become more complicated. To develop its user networks, Facebook began to offer "open to all" options for users. These allowed users to choose specific individuals who could search their own personal profiles by using Google.

In 2006, Facebook added a further "news feed" feature (Boyd and Hargittai 2010) that notified users of any movements made by users' friends. This news feed was criticized because it automatically sent others' updates by the SNS, rather than allowing users to take the initiative of viewing others' updates (Boyd 2008). An "anti-Facebook" movement developed in response to this new feature, which forced Facebook to add an option for users to choose whether they wanted to let their posts be included in the news feed service (Meredith 2006).

These developments in Facebook's privacy setting options have developed along with various new features that have been introduced. However, sometimes the changes in features take place so quickly that users are unable to make use of the latest privacy setting options.

In 2010, Facebook introduced a revolutionary feature, allowing third parties to link and share personal data on Facebook. For example, by adding a Facebook "like" button to websites, if Facebook users clicked on the button, content on websites would be shown on users' Facebook timeline wall when users logged into their accounts. The new feature allowed Facebook to link content from external websites, but this created more serious privacy concerns. In 2010, privacy protection organizations submitted their complaints to the Federal Trade Commission, arguing that Facebook's privacy settings were far too complicated for even experienced users to understand (Cheng 2009). Similar complaints which accused Facebook of privacy infringements appeared in Canada (McCarthy 2010). Mark Zuckerberg, the founder of Facebook, responded by saying that while he was still at Harvard, he was widely quoted by Internet users (Orlowski 2010), and Facebook became a target for debates concerning privacy protection.

Notably, there are two types of personal data that can be disclosed on Facebook: personal information provided by users, such as personal profiles or posts published by users on their timeline wall, and users' interactions with other users, such as by requesting others to be "friends" or being other users' friends. Making friends on Facebook allows users to interact with each other, which is the main function of SNSs but also creates difficulties in terms of privacy protection owing to the fact that when users become friends, they can view each other's personal files and posts published on timeline walls.

As the number of Facebook users grows, the quantity of personal data circulated on Facebook has grown substantially, thus exacerbating privacy concerns. At the forefront of these privacy concerns is the role of the Facebook service operator (Facebook Help Center 2011), who is able to determine how users' data are to be preserved and managed. Nonetheless, apart from Facebook itself being the potential privacy infringer of users as a service provider, it is often ignored that Facebook users themselves also have great potential to infringe on other users' privacy.

The privacy options of Facebook allow users to choose who can and who cannot see their personal profile posts (Facebook Help Center 2014a). The available options provided include "public," "friends," "friends of friends," "custom," and "only the user" (Facebook Help Center 2014b). If users choose the "public" option, then anyone including people who are not Facebook users can see content. If users choose "friends of friends," then users' friends and any friends of their friends can see posts. When users choose "friends" (or friends of anyone tagged), then users post to users' friends on Facebook; if anyone else is tagged in the post, then the audience expands to include the tagged people and their friends. When users choose "only the user," that means the posts are only visible to the user. Lastly, if users choose "custom," then users can selectively share something with specific people or hide it from specific people.

Users may think they can control who can view their posts on their timeline by controlling who can be their friends; however, because Facebook provides privacy options such as "friends of friends" and "public," in some cases users can see or share others' posts without being friends with each other. For example, users A and B can see each other's posts if A and B have friends in common and if A and B both choose the "friends of friends" option while publishing posts. If A or B choose the "public" option, then A and B can see public posts by accessing each other's timeline.

In order to encourage more people to use Facebook, the default setting on Facebook user profiles is "public," and users can choose from "public" to "friends of friends." The default setting of posts had been "public" until May 2014 when Facebook changed it to "friends" (Magid 2014). The result is that without being "friends," users can track others' public personal profiles through the Internet or see each other's public posts without having friends in common.

Choice of privacy settings on Facebook

According to a survey of student Facebook users conducted in 2008, although Facebook provides several privacy setting options, 87 percent of student users would not alter the default setting provided by Facebook (Strater and Lipford 2008). More importantly, although many users have been aware of privacy concerns when using SNSs, this has not led them to change the default privacy settings. This may be due to Facebook users not being aware of how their personal information may be disclosed including contact phone numbers or addresses in personal profiles, leading to users becoming victims of stalking or identity theft. On the basis of Strater and Lipford's (2008) psychological research, there are two reasons why users of SNSs may not select high privacy protection options.

The first reason relates to interactive options that impact on users' privacy settings. Many users log on to Facebook to make new friends or post new updates and tend to forget the existence of personal profiles. Nonetheless, the default setting of personal profiles, as noted above, is "public," which can lead to personal details contained in profiles being misused. In addition, when users publish new posts on their timelines, these are usually only limited to friends who may choose to "like" posts they are interested in. It is apparent, however, that many users are unaware that their posts may be viewed by much larger numbers of people than the few individuals who choose to "like" them or provide comments. For example, user A may invite friends to join a meeting and leave friend B's contact phone number in a post or comment, and yet B may not have disclosed his own contact number in his own personal profile. As a result, all A's friends (who may not necessarily be B's friends) would be able to see B's contact number from accessing A's post. In brief, users tend to ignore the real impact of publishing a post because most of their attention is directed to those whom they interact with directly, forgetting that others will see the post too.

The second reason for not choosing high privacy protection options is that the privacy setting options of Facebook are difficult for users to understand (Strater and Lipford 2008, s. 4.3.2). While users need to learn about user privacy options, under pressure from the privacy protection movement, Facebook has changed its privacy option models on a frequent basis, which has made the relationship between Facebook and privacy even more complicated (Fung 2014). In addition, Facebook often provides many interesting apps and games for users to download so as to encourage users to interact with each other. However, these apps and games may require users to accept specified privacy terms, and users may not be aware of changes being made to their privacy settings as a result of downloading games on Facebook. It seems, therefore, that there are many privacy traps on Facebook that users often ignore.

According to research conducted by Boyd and Hargittai (2010) in which Facebook user behavior in 2009 and 2010 were compared, it was found that Facebook users visited privacy pages more frequently, which is an indication that users may be valuing privacy protection more than before. Nonetheless, it takes time to educate users to fully understand all of the privacy options provided on the Facebook platform.

The concept of privacy in social networking sites

The evolution of privacy in the digital age

The rise of social networking sites such as Facebook and MySpace has changed society's definition of the Internet (Cassidy 2006). SNSs are popular not only because users can establish networks or communities but because they enable users to express themselves and establish self-identities (Cassidy 2006). Although scholars may have different definitions on "privacy," it is widely accepted that the justification for privacy protection is to allow individuals to decide how to define themselves and control personal information. As a result, privacy protection enables individuals to express emotions and to engage in self-evaluation and self-determination and encourages autonomy (Rempell 2006).

Because the Internet allows information to be instantly accessible to billions of users, this has created an unprecedented challenge for privacy protection. Renowned privacy scholar Solove (2007) has suggested that the traditional concept of privacy should be expanded to acknowledge that individuals are less likely to be able to control personal information in the digital age. Solove (2007) has pointed out that privacy regulations should correspond to individuals' expectation of confidentiality and oblige others to meet such expectations.

Despite the difficulties of establishing a precise standard of privacy in the digital age, it is worth considering the way in which protection of privacy has evolved in the US courts. In the case *Lawrence v Texas*, for example, the Supreme Court overruled its previous decision and struck down a sodomy law in Texas on the ground of constitutional protection of the privacy of sexual relations (see Millier 2009, p. 550).

In the case of *United States v Gines-Perex*, a police officer downloaded a suspect's photo from a private website and identified the suspect on the basis of the photo. The suspect argued that the photo was protected by privacy laws, thus making the officer's behavior illegal (Millier 2009, p. 551). In its decision, the court established a two-step standard for determining the boundaries of constitutional privacy protection on the Internet. First, the court should examine whether the individual has an expectation of a right to privacy, and secondly, the court must determine whether or not that expectation was reasonable. In the case before the court, it was decided that because any information on the Internet is freely available for others to access, the court found there to be no reasonable expectation of privacy in information placed on the Internet.

In another case, *Sanders v American Broadcasting Companies* (1999), the plaintiff was a reporter and worked undercover at a tele-psychic company. She videotaped interactions she had with an employee of the company, who later sued her and her employer for invasion of privacy. The main issue of this case was whether an employee could have a reasonable expectation of privacy in a working environment in which conversations might be overheard by other co-workers. The court ruled that, taking into account the nature of the conduct and surrounding circumstances, an employee has a legitimate expectation that his conversations will not be secretly videotaped even though such conversations may not have been completely private. This case has established that privacy protection can be possible in an open environment as long as the individual has a reasonable expectation that his or her privacy will be respected.

The privacy illusion of Facebook users

Many Facebook users believe that all the information disclosed on their account pages can only be accessed by "friends." According to Sophos' survey in 2008, 41 percent of Facebook users would give emails, birthdays, and even phone numbers to a stranger (Sophos 2008). In 2007, Miss New Jersey USA Amy Polumbo's Facebook account was compromised and her half-nude photos were released to the media (Parry 2007). Polumbo confirmed that she was blackmailed because of the photos, which she supposed had only been circulated among her Facebook friends.

Another privacy illusion of Facebook is that even if privacy settings have been chosen widely, personal information may still be insecure. For example, by clicking "Yes" on the terms and conditions of Facebook, users allow Facebook to mine data generated by themselves (Millier 2009, p. 546). All the information provided in users' profiles and timelines can be used by Facebook to create customized advertisements. At present this remains legal in view of the users' consent when creating an account (Klosek et al. 2008). In 2008, the Canadian Privacy Commission dealt with a complaint against Facebook which accused it of not performing its duty to inform users how their personal information would be disclosed to third parties for advertising and commercial purposes under the Canadian Personal Information Protection and Electronic Documents Act (PIPEDA) (Canadian Internet Policy and Public Interest Clinic 2008). The Assistant Privacy Commissioner found that the majority of the complaints were well-founded. Although Facebook agreed to comply with the ruling and made sweeping changes to its privacy settings, it failed to meet the standards set out in PIPEDA. In April and May of 2010, Facebook responded further by improving ease of access to privacy settings on its site and improving "transparency surrounding what user information application and website advertisers/developers can access when a user interacts with their services." According to the Canadian Internet Policy and Public Interest Clinic (2015), Facebook still has failed to address the core concerns raised by the operation of its site and as such remains in violation of PIPEDA:

> In particular, Facebook continues to pre-select default settings for its users that do not reflect reasonable expectations or the sensitivity of the information in question. Also, it does not appear that Facebook intends to provide granular control over data provided to advertiser/ developers (Canadian Internet Policy and Public Interest Clinic 2015).

In 2007, Google proposed a merger with the DoubleClick corporation, leading to investigation both by the US Federal Trade Commission and the European Commission (Kawamoto and Broacher 2007; Kawamoto 2008). The merger would have enabled users' Internet search data collected by Google to be analyzed in conjunction with users' preference data collected by DoubleClick. Although the merger was approved by both the FTC and the EC, it raises a number of concerns about corporate access to, and use of, user-generated data with the potential to infringe privacy.

To conclude, it is clear that although most of the content on SNSs is generated by users (user-generated content), users often ignore privacy protection options when engaging in social networking and sometimes invade other users' privacy inadvertently. In view of the massive quantities of data that are present on SNSs, the potential infringement of privacy is extreme, both by social networking sites as well as by users themselves. In order to address these concerns, it is necessary to examine current legal and regulatory frameworks that govern social networking.

Privacy protection regulations in the world of social networking

United States of America

Section 230 of the Communications Decency Act 1996 (CDA) in the United States provides general protection for SNSs in respect of content published on their platforms by users or by third parties. Section 230 states:

> No provider or user of an interactive computer service shall be treated as the publisher or speaker of any information provided by another information content provider, namely online intermediaries that host or republish speech are protected against a range of laws that might otherwise be used to hold them legally responsible for what others say and do (s. 230 Communications Decency Act 1996).

This creates a safe haven for YouTube users to upload their own videos, Amazon to offer countless user reviews, and Facebook to offer social networking to hundreds of millions of Internet users. Under the protection of CDA, Facebook would not be held liable for content published on its platform by its users or third parties.

According to Millier (2009, p. 553), the courts have applied a broad interpretation to the protection afforded by the CDA and have not obliged Internet service providers to perform active content filtering. However, in 2008, the Ninth Circuit Court redefined the scope of the CDA protections in the case *Fair Housing Council v Roommates.com* (2008). In this case, the defendant, Roommates.com, required users to provide answers to a set of questions that included users' preferences in sex, sexual orientation, and family status for future roommates, allegedly violating the Fair Housing Act by reason of illegal discrimination against housing. The Court found that by requiring subscribers to provide the unlawful preference information as a condition of accessing its service, and by providing a limited set of prepopulated answers, Roommates.com became much more than a passive transmitter of information provided by others, but rather became the developer, at least in part, of that information.

The Court reversed the decision of the District Court and held that Roommate.com was not immune under section 230 of the Communications Decency Act 1996 (*Fair Housing Council v Roommates.com*, p. 979). The Court further pointed out that Roommates.com was protected by the CDA for publishing the "additional comments" section but not for 1) posting questionnaires that required disclosure of sex, sexual orientation, and familial status; 2) limiting the scope of searches by users' preferences on a roommate's sex, sexual orientation, and familial status; and 3) providing a matching system that paired users based on those preferences (*Fair Housing Council v Roommates. com*, p. 979).

The legal reasoning of the Circuit Court in this case may lay a legal foundation for requiring Facebook to better protect users' privacy. Because it can be argued that by designing an Internet platform and encouraging users to publish and share information, Facebook may be more than a passive transmitter of content and will not be protected under the terms of the CDA.

European Union

The European Union (EU) adopted the Data Protection Directive in 1995 (Europa 1995) that confirmed privacy as a fundamental human right and determined that personal data collection should be done for legal purposes and under legal process. More importantly, the Directive provided an "opt in" option that required individuals to provide consent (to opt in) before SNSs could collect and use personal data. Many EU member states have proposed their own data protection laws based on the principles laid out by this Directive. For example, the UK passed its Data Protection Act (1998), which has given personal data a broad definition and prohibited the transfer of personal data to other countries that fail to reach the same levels of protection required by the EU Directive.

Although the Directive has established a good level of protection of personal privacy, it did not envisage or consider the rapid development of SNSs when it was created in 1995. It is questionable whether social networking service providers and users are subject to the provisions of the Directive. For example, Wong (2010, p. 131) questions whether Facebook can be considered to be a data controller and whether Facebook users are data subjects under Article 2 of the Directive. Arguably, SNS service providers should be defined as data controllers as they provide a service platform for users to link and interact with each other and because they clearly specify the purpose and methods used to process the personal data they collect.

Van Alsenoyn et al. (2009) argue that users can be defined as data subjects because they are able to choose the extent to which personal data may be disclosed as well as the methods used to disclose data (by texts or photos). Notably, in the case of users, the Directive may not apply if data are processed by natural persons in the course of a purely personal or household activity. The European Court of Justice has ruled that the information published on the Internet would not be exempted under this Article because publicly accessible information could be accessed by unlimited numbers of people and would not be considered as purely personal or household activity (ECJ 2004a, 2004b).

Arguably, therefore, the exemptions may not be applicable to SNS users. The opinion documents which further clarified the application of Article 29 of the Directive (Europa 2009) suggested that the question of whether SNS users can be data controllers under the Directive depends on whether user accounts are set to "public" or "private." If the personal profile of the user can be accessed by all users of the social networking site and can be found through Internet search engines, it is not a private account. In conclusion, if users are identified as data controllers, then the personal and household exemption would not apply, and they would be required to obey all rules stipulated in the Directive. In other words, if users are identified as data controllers, they need to seek the consent of third parties when they wish to use third parties' personal data. For example, if data controller users wish to upload photos of friends and use tagging functions, they would need to seek approval from friends to upload photos and then seek the friends' consent again to tag them.

Personal Information Protection Act (Taiwan)

In Taiwan, a reformed Personal Information Protection Act (2010a) was passed in 2010. According to the Act, in theory, social networking service providers and users are subject to the provisions of the Act. Article 19 of the Act stipulates the legal process for non-government agencies to follow when collecting personal information. This requires agencies to inform the data subject and obtain informed consent prior to data collection unless the data are already made public on the Internet.

The first paragraph of Article 51 states:

> The provisions of this Law are not applicable to the following situations: When an individual who collects, processes or uses personal information in the course of personal activity of a domestic nature.

According to a press release by the Ministry of Justice (2010b), this exemption can be applied to Facebook users who publish posts or photos of social and household activities. The legal principles laid out in the new reformed Act in Taiwan seem to be close to the EU Data Protection Directive.

To summarize, social networking site providers and users are all subject to the regulations of the Act, unless exempted under the social and household doctrine. Therefore, the existing interpretations of EU Directive and relevant rulings made by the European Court of Justice can be useful reference materials for Taiwan to apply when applying this Act to privacy issues in the future.

Conclusions

Because of the predetermined user privacy settings that are used on SNSs, many users may not be fully aware of the extent to which their privacy could be infringed while using sites. If users usually are not fully aware of the privacy setting options they have, it seems unreasonable to expect them to choose the most suitable privacy option available. On the basis of the legal and policy frameworks developed in the US and EU regarding privacy protection on the Internet, social networking service providers have a degree of responsibility for safeguarding the privacy of SNS users. Arguably the best way in which the privacy of SNS users can be protected is for users to be educated more effectively about the privacy setting options that are available and how these would impact on their and even their friends' personal data. It is reasonable to require social networking service providers such as Facebook to offer a default privacy setting that most protects its users.

Bibliography

Boyd, D. M. (2008) "Facebook's Privacy Trainwreck: Exposure, Invasion, and Social Convergence," *Convergence* 14: 13–20.

Boyd, D. M. and Hargittai, E. (2010) "Facebook Privacy Settings: Who Cares?," *First Monday* 15(8), http://www.uic.edu/htbin/cgiwrap/bin/ojs/index.php/fm/article/view/3086/25890 (accessed 2 August 2010).

Boyd, D. M. and Ellison, N. (2007) "Social Network Sites: Definition, History, and Scholarship," *Journal of Computer-Mediated Communication* 13: 210–30.

Canadian Internet Policy and Public Interest Clinic (2015) "Complaint Against Facebook 2008–2010," https://cippic.ca/en/Facebook (accessed 24 February 2015).

Canadian Internet Policy and Public Interest Clinic (2008) "CIPPIC Files Privacy Complaint Against Facebook," https://cippic.ca/en/news/339/54/uploads/NewsRelease_30May08.pdf (accessed 30 May 2008).

Cassidy, J. (2006) "ME Media: How Hanging Out on the Internet Became Big Business," *The New Yorker*, 15 May, http://www.newyorker.com/magazine/2006/05/15/me-media (accessed 24 February 2015).

Chan, J., Goggin, G. and Bruce, J. (2010) "Internet Technologies and Criminal Justice," in *Handbook of Internet Crime*, Y. Jewkes and M. Yar (eds.), Cullompton, UK: Willan Publishing, 582–602.

Cheng, J. (2009) "FTC Complaint Says Facebook's Privacy Changes Are Deceptive," *Ars Technica*, 21 December, http://arstechnica.com/tech-policy/news/2009/12/ftc-complaint-says-facebooks-privacy-changes-are-deceptive.ars (accessed 24 February 2015).

Communications Decency Act, 47 U.S.C. § 230(c).

Data Protection Act (1998) *UK legislation*, http://www.legislation.gov.uk/ukpga/1998/29/contents (accessed 24 February 2015).

European Court of Justice (ECJ) (2004a), C-101/01, Bodil Lindqvist, European Court of Justice, Luxembourg, www.curia.europa.eu (accessed 24 February 2015).

European Court of Justice (ECJ) (2004b), O.J. C 7/3, European Court of Justice, Luxembourg, www.curia.europa.eu.

Europa (1995) "Data Protection Directive 95/46/EC (DPD)," 23 November, http://eur-lex.europa.eu/LexUriServ/LexUriServ.do?uri=CELEX:31995L0046:en:HTML (accessed 24 February 2015).

Europa (2009) Article 29 Data Protection Working Party, "Opinion 5/2009 on Online Social Networking (WP 163)," 12 June, ec.europa.eu/justice/policies/privacy/docs/wpdocs/2009/wp163_en.pdf (accessed 24 February 2015).

European Parliament (1995) "Directive 95/46/EC of the European Parliament and of the Council of 24 October 1995 on the protection of individuals with regard to the processing of personal data and on the free movement of such data," http://eur-lex.europa.eu/legal-content/en/TXT/?uri=CELEX:31995L0046 (accessed 24 February 2015).

Electronic Frontier Foundation (2015) "Section 230 of the Communications Decency Act," https://www.eff.org/issues/cda230 (accessed 24 February 2015).

Facebook Help Center (2014a) "Basic Privacy Settings & Tools," https://www.facebook.com/help/325807937506242/ (accessed 15 December 2014).

Facebook Help Center (2014b) "Basic Privacy Settings & Tools," https://www.facebook.com/help/211513702214269 (accessed 6 February 2015).

Facebook Policy (2011) Sections: "Information we receive" and "Information we collect," http://www.facebook.com/policy.php (accessed 30 August 2011).

Fair Hous. Council v. Roommates.com, 521 F3d 1157 (9th Cir. 2008), http://cdn.ca9.uscourts.gov/datastore/opinions/2012/02/02/09-55272.pdf (accessed 24 February 2015).

Fung, B. (2014) "Your Facebook Privacy Settings Are About to Change. Again," *Washington Post*, 8 April, http://www.washingtonpost.com/blogs/the-switch/wp/2014/04/08/your-facebook-privacy-settings-are-about-to-change-again/ (accessed 24 February 2015).

Grimmelmann, J. (2009) "Facebook and the Social Dynamics of Privacy," *Iowa Law Review* 95: 1–52.

Hargittai, E. (2007) "Whose Space? Differences Among Users and Non–users of Social Network Sites," *Journal of Computer–Mediated Communication* 13: 276–97.

Kawamoto, D. and Broacher, A. (2007) "FTC Allows Google-DoubleClick Merger to Proceed," *CNET News*, 20 December, http://news.cnet.com/FTC-allows-Google-DoubleClick-merger-to-proceed/2100-1024_3-6223631.html (accessed 24 February 2015).

Kawamoto, D. (2008) "With Europe's OK, Google Closes DoubleClick Acquisition," *CNET News*, 11 March, http://news.cnet.com/8301-10784_3-9890858-7.html (accessed 24 February 2015).

Klosek, J., Beliveau, N., Bleier, L., Kumar, N., Lee, Y. and Shreve, J. (2008) "Information Services, Technology and Data Protection," *International Lawyer* 42: 623.

Magid, L. (2014) "Facebook Changes New User Default Privacy Setting to Friends Only – Adds Privacy Checkup," *Forbes*, 22 May, http://www.forbes.com/sites/larrymagid/2014/05/22/facebook-changes-default-privacy-setting-for-new-users/ (accessed 24 February 2015).

McCarthy, C. (2010) "Toronto Law Firm Preps Facebook Privacy Suit," *CNET News*, 8 July, http://news.cnet.com/8301-13577_3-20009956-36.html (accessed 24 February 2015).

Meredith, P. (2006) "Facebook and the Politics of Privacy," *Mother Jones*, 14 September, http://motherjones.com/politics/2006/09/facebook-and-politics-privacy (accessed 24 February 2015).

Millier, S. (2009) "The Facebook Frontier: Responding to the Changing Face of Privacy on the Internet," *Kentucky Law Journal* 97: 541–64.

Orlowski, A. (2010) "Facebook Founder Called Trusting Users Dumbf*cks," *The Register*, 14 May, http://www.theregister.co.uk/2010/05/14/facebook_trust_dumb/ (accessed 24 February 2015).

Parry, W. (2007) "'Private' Online Photos Really Aren't," *USA Today*, 12 July, http://usatoday30.usatoday.com/life/television/2007-07-12-421814674_x.htm (accessed 24 February 2015).

Rempell, S. (2006) "Privacy, Personal Data and Subject Access Rights in the European Data Directive and Implementing UK Statute: Durant v. Financial Services

Authority as a Paradigm of Data Protection Nuances and Emerging Dilemmas," *Florida Journal of International Law* 18: 808–31.

Sanders v American Broadcasting Company, 978 P.2d 67, 70 (Cal. 1999).

Solove, D. J. (2007) *The Future of Reputation: Gossip, Rumor, and Privacy on the Internet*, New Haven, CT: Yale University Press.

Sophos (2008) "Facebook: The Privacy and Productivity Challenge," http://www.sophos.com/security/topic/facebook.html (accessed 6 February 2015).

Strater, K. and Lipford, H. R. (2008) "Strategies and Struggles with Privacy in an Online Social Networking Community," in *Proceeding BCS-HCI '08 Proceedings of the 22nd British HCI Group Annual Conference on People and Computers: Culture, Creativity, Interaction* vol. 1, Liverpool, UK: British Computer Society, 112, http://dl.acm.org/citation.cfm?id=1531530 (accessed 24 February 2015).

Taiwan Ministry of Justice (2010a) "Personal Information Protection Act," 26 May, http://law.moj.gov.tw/Eng/LawClass/LawAll.aspx?PCode=I0050021 (accessed 24 February 2015).

Taiwan Ministry of Justice (2010b) "Third reading by the Legislative Yuan: Personal Data Protection Act," 27 April, http://www.moj.gov.tw/public/Data/0427164423187.pdf (in Chinese) (accessed 24 February 2015).

Van Alsenoy, B., Ballet, J., Kuczerawy, A. and Dumortier, J. (2009) "Social Networks and Web 2.0: Are Users Also Bound by Data Protection Regulations?," *Identity in the Information Society* 2: 65–79.

Wong, R. (2010) "Data Protection: The Challenges Facing Social Networking," *Brigham International Law and Management Review* 6: 127–52.

16

An Introduction to Cyber Crowdsourcing (Human Flesh Search) in the Greater China Region

Lennon Y. C. Chang and Andy K. H. Leung

Introduction

Cyber crowdsourcing – known in Chinese as *renrou sousou*, literally "human flesh search" (HFS) – is a type of collective online action aimed at discovering facts relating to certain events and/or publicizing details concerning a targeted individual (Cheung 2009; Herold 2011; Ong 2012). It involves the tracking down and publishing on the Internet of information that might help to solve a crime or to disclose personal information of someone who has allegedly engaged in corrupt practices or immoral behavior (Hatton 2014; Ong 2012). It can also be used to identify people in events that attract the public's attention, such as love affairs of celebrities. Examples from the Chinese online community include news about people who abuse animals, teenagers who do not respect their elders, wealthy children who do not care about the feelings of others, and the behavior of corrupt officials. Although some Chinese researchers claim it is an online phenomenon unique to the Greater China region (Cheng and Xue 2011), similar cases have emerged in the West in recent years. One recent example is the identification of suspects responsible for the Boston marathon bombings (Wadhwa 2013).

HFS has become a powerful tool used by netizens concerned about social issues in the Greater China region, including the People's Republic of China (China), Hong Kong, and Republic of China (Taiwan). Recalling Peter Steiner's famous cartoon in *The New Yorker* in 1993 that "On the Internet, nobody knows you're are a dog" (Steiner 1993), through cyber crowdsourcing, netizens can not only find out that you are a dog, they will even know which type of dog you are. Originally, HFS involved only a small group of people seeking information relating to certain incidents, but it has since become more like a manhunt. The large number of netizens involved has transformed it into a tool capable of blaming and punishing behavior considered socially immoral, uncivilized, or even illegal. HFS is based on the idea that once a person has used the Internet, he or she must have left some trace of evidence for others to uncover. Netizens who undertake cyber crowdsourcing have been described as the most talented detectives and better than the FBI when it comes to tracing offenders (Xiao 2011; Yang and Zhang 2010).

Mobile apps, Internet forums, and other online community sites such as Bulletin Board Systems (BBS), provide excellent platforms for cyber crowdsourcing. According to Internet World Stats – the website of Internet market research company Miniwatts Marketing Group – there are more than 650 million Internet users in the Greater China region, most of whom are in online communities such as Golden (HK), HK Discuss (HK), U-wants (HK), PPT (Taiwan), Weibo (China), and Facebook (HK and Taiwan). An online advertising research company has reported that these Chinese BBS online communities had nearly 43.8 million visitors in 2009, with numbers increasing since then (Fu 2009). China Internet Watch (2014) reported that there were more than 167 million monthly active users on Weibo in 2013. In Taiwan and Hong Kong, 70 percent of Internet users are on Facebook. Other online communities such as Golden (Hong Kong) and PTT (Taiwan) are also very popular in Hong Kong and Taiwan. These substantial numbers of netizens and platforms provide a good base for cyber crowdsourcing activities in the Greater China region.

This chapter provides an insight into cyber crowdsourcing in the Greater China region. It will, firstly, define and then describe the origins of cyber crowdsourcing. Based on archival studies conducted in 2012, it will categorize cyber crowdsourcing cases between 2003 and 2012. It will then discuss the positive and negative impacts of cyber crowdsourcing and discuss the possibility of using cyber crowdsourcing in criminal investigations.

What is human flesh search?

Human flesh search (HFS) has been defined as any kind of message that appears in a question-and-answer format targeted at obtaining knowledge (Liu and Hao 2009). However, when we investigate cases of

HFS, they all relate to crowdsourcing of particular persons. That's why the term "human flesh" is used to describe cyber crowdsourcing in Chinese.

HFS relies on social networking platforms in which netizens (individuals actively involved in the online community) can provide information and clues. Their fellow netizens may then conduct further investigations to uncover more information based on the initial information and clues provided. These platforms thus differ from the automated platforms offered by conventional search engines (Wang et al. 2010) such as Google and Baidu, which involve one-way communication in which users retrieve information from an archived database of information collected by a web-crawler. In the case of conventional search engines, if the information requested does not exist in the database, then a negative response will be provided (Liu 2009).

Although HFS relies heavily on traditional search engines to find information, an Internet user can also obtain information through contributions from other netizens and then engage in further searching based on the information accumulated. It thus constitutes a form of multiway communication and has been called "people-powered search" (Wang et al. 2010, p. 45). Those accessing the platform can ask specific questions, and others can contribute information or clues based on their existing knowledge. These platforms thus facilitate mass cooperation with thousands of people using every possible method to help find the answer to the question posed (Wang 2009; Lu 2012).

Selected cases of human flesh search in the Greater China region

Du Yao

The first suspected case of cyber crowdsourcing in China occurred in early 2006 when a number of netizens decided on their own initiative to establish the identity of an anonymous online celebrity nicknamed "Du Yao" (Poison), an online game player who was having extramarital affairs with other game players. Eventually, all known information about him, including his educational and family background and previous girlfriends, as well as photographs and even images of his payslips, were posted on the celebrity board of the Tianya forum (www.tianya. cn). At its peak, posts related to Du Yao were receiving 2,000 to 3,000 replies. In mid-2006, a comment concerning loss of privacy in the digital age was published by the Xinhua news website, which noted that in cyberspace it was no longer the case that "nobody knows you are a dog" but that "everyone knows you are a dog" (Ju 2006).

Wang Fei

Another famous case, widely accepted as one of the first cases of HFS, involved Wang Fei, a woman who committed suicide in 2007 because her husband cheated on her. Prior to taking her own life, she wrote a blog about her situation. This motivated netizens to take matters into their own hands, and they set about revealing personal information relating to her husband, including his workplace, home address, and telephone number. Some netizens started to harass him by phoning him at his home and his workplace. He was dismissed by his employer, and his family were harassed. Eventually, he successfully sued three website administrators (Tang 2010; Ye and Li 2009). It was the first case of this nature to be brought to court, and it created a precedent in China, establishing that social forum websites are responsible for monitoring HFS activities.

Government official molestation

In 2008, a video entitled "Government Official Molestation" was disseminated over the Internet in China. The video showed a confrontation between the parents of an 11-year-old girl and a mayor-rank official accused of molesting the girl. In the video, the man said, "I am a government official and have the same rank as a city mayor. There is no way you can sue me!" The government official's haughty statement outraged netizens, who began connecting and communicating with one another to reveal personal and other information about the official. Two days later, Jiaxiang Lin, Deputy Director-General and Party Secretary of Shenzhen Maritime Affairs Bureau (SMAB) under the Ministry of Transport (MOT), was singled out as the perpetrator. In the meantime, personal information about Lin, such as his birthplace, mobile phone numbers, and alma mater were disclosed by eager netizens over the Internet. As a result, MOT announced that Lin had been removed from all administrative responsibilities and party duties and was required to make a public apology for his misbehavior (People 2008).

Edison Chan

In 2008, a netizen released a series of photos of a naked female and male who looked like the famous Hong Kong pop stars Gillian Chung and Edison Chan onto the adult photo area of a Hong Kong online forum. The post was deleted a few hours later, but netizens had downloaded the photos and reuploaded them onto a variety of online forums. Further photographs of naked women were uploaded, and five more female celebrities were identified by netizens. After investigation by the police, it was discovered that the photographs had been copied by a computer repair shop employee when Edison Chan brought his notebook in for fixing (Reuters 2008).

Cat abuse case

Another sensational case concerning "Cat Abuse in Shun Tin Village" attracted significant public attention in Hong Kong. The case involved five teenagers who allegedly repeatedly kicked a cat. The cat sustained severe injuries including organ failure and was subsequently euthanized. When this incident was reported in the media, the public held an Internet trial and tracked down the alleged abusers. Within a short period of time, five teenagers were identified and their personal information was posted on the Internet. Thousands of criticisms and insulting messages directed at the abusers were posted. The teenagers claimed they were innocent and accused netizens of making mistakes in their online searching. They asked the police to investigate in a bid to establish their innocence. When the teenagers were eventually found to be innocent, some netizens suspected that false information was uploaded by someone who knew the teenagers and intended to harm them (AppleDaily 2012).

Middle-finger Hsiao

In 2010 in Taiwan, a photograph was posted online showing a young driver refusing to give way to an ambulance and showing his middle finger (a sign of foul language) to the ambulance driver. The old lady in the ambulance died because of a delay in obtaining medical treatment. The media reported this incident and immediately triggered HFS. A PhD student named Hsiao, then a PhD student at a national university in Taiwan, was identified, and his name and other personal information were revealed. He was expelled from the university (Taipei Times 2010).

The prevalence of HFS in the Greater China region

The cyber crowdsourcing phenomenon that has emerged is particularly prevalent in Chinese societies. Wang et al. (2010) reported that, of the 35 cases they recorded from 2001 to 2007, 31 occurred within the Greater China region. Furthermore, from January to May 2010, they recorded 87 cases of cyber vigilantism, of which 85 took place in the Greater China region. Most of these cases were concerned with the pursuit of truth or justice.

A Wisenews search carried out by the authors using the keywords "ren-rou sou-sou" for the ten-year period 1 January 2003 to 31 December 2012 identified 363 cases with 13,914 news articles on cyber crowdsourcing (see Table 16.1 for details). These cases can be categorized into seven main types: misconduct by government officials; maintenance of social order; appreciation of good deeds; misunderstanding or misuse of HFS; entertainment including curiosity about certain incidents; expression of negative emotions; and police-netizen cooperation.

In Taiwan, there have been no known cases of HFS targeting government officials. Most cases (87%) related to minor crimes such as animal abuse or theft or immoral incidents such as not respecting the elderly or relationship affairs. However, in Hong Kong less than half of the cases (46%) related to minor crimes or immoral incidents. More than 30 percent of the cases in Hong Kong related to entertainment, involving personal affairs and sexual relationships, sex videos taken between

Table 16.1 Category of human flesh searching cases in the Greater China region

Case Category	Mainland China		Hong Kong		Taiwan	
	n	%	n	%	n	%
1. Misconduct of government (punishing)	39	15	1	8	0	0
(a) Corruption activities or abuse of power	25	10	0	0	0	0
(b) Illegal activities by the second generation of government officials	7	3	0	0	0	0
(c) Sex scandals of government officials	7	3	1	8	0	0
2. Crime control and social control (punishing)	140	53	6	46	53	87
(a) Minor crime issue	101	39	3	23	40	66
(b) Immoral activities	39	15	3	23	13	21
3. Appreciating good deeds	32	12	0	0	5	8
(a) Saving others	5	2	0	0	2	3
(b) Finding people	18	7	0	0	3	5
(c) Finding good people who help others anonymously	9	3	0	0	0	0
4. Misunderstanding and misuse of HFS	8	3	0	0	0	0
5. Entertainment	24	9	4	31	1	2
(a) Sex scandals or negative news of artists	2	1	2	15	1	2
(b) Incidents about business activities	0	0	1	8	0	0
(c) Curiosity about certain incidents	22	8	1	8	1	2
6. Expression of negative emotions	10	4	2	15	0	0
7. Police-netizen cooperation	9	3	0	0	1	2
Total	262	100	24	100	61	100

lovers, or sex scandals involving popular figures. It is worth nothing that Hong Kong's Umbrella Movement that happened in late 2014 is likely to result in an increase in the number of cases relating to the misconduct of government officials (in particular, allegations of brutality of the police against demonstrators).

The situation in Mainland China is different from that in Taiwan and Hong Kong. In Mainland China, HFS cases mainly involve targeting government officials, maintaining social order, and appreciating good deeds. Among the 262 cases identified in Mainland China, more than half (53%) are related to crime control and social control issues. Since netizens believe that the police will ignore these cases or will handle them inadequately or unfairly, they tend to solve the cases by themselves. Apart from the cases relating to minor crime or deviance, netizens in Mainland China have also successfully identified a number of corrupt government officials through HFS. Most of these HFS cases have either forced the Chinese government to take, or have supported them in taking, action to deal with identified cases of corruption. Netizens treat HFS as a weapon to discover counterfeit products, to uncover the illegal actions of government officials, and to reveal the truth. Through HFS, people are more willing to contribute to the investigation as they are less worried about their identity being revealed. They appreciate good deeds done and promote incidents of helping others that they think match traditional virtues of Confucianism (Wang 2009; Lu 2012; Xu and Li 2011).

China is undergoing rapid and significant economic development. At the same time, the country is facing a range of challenges including misconduct and corruption among public officials. Researchers have argued that HFS could lead to the public surveillance of government officials and anti-corruption campaigns by netizens (Liu and Zhuang 2012; Shu 2012). As the public has little confidence in law enforcement agencies, HFS has assumed importance in addressing misconduct and corruption among officials. HFS can also help to fight crimes involving government officials (Liu 2011). HFS enables netizens to challenge the power and privilege of government officials (Chen 2009).

Ruan (2011) pointed out that HFS could help address governance issues and the inadequacies of law enforcement and thereby maintain social justice in China. HFS could also increase social effectiveness. In the face of bureaucratic and inefficient government agencies and officials, netizens can take action themselves to deal informally with crimes that are often ignored, either purposely or unconsciously, using HFS. Although HFS is not organized or planned in advance, crowdsourcing can run very smoothly and appear to be organized. Through HFS, clues relating to an event can be found and analyzed quickly as the information is provided collectively by netizens. Although these netizens have no prior training as investigators, the quality of the information or evidence collected is usually as good as that collected by professional investigators (Ye and Li 2009; Liu 2009).

HFS can also help to reveal the truth to the public. In China, the media is monitored and news is filtered by the government. With the use of HFS and some simple techniques such as using abbreviations or symbols, the public can access news that has not been censored. Yuan and Zhao (2012) argued that this will challenge China's current closed governance style and lead to openness and transparency of information in China.

Although HFS can provide benefits and opportunities, there are also negative impacts that should not be ignored. Most netizens participate in HFS voluntarily because they think it is their responsibility to contribute to the maintenance of social justice (Liu and Hao 2009). HFS sometimes involves a black-and-white moral judgment without any concern for the underlying details of the incident (Chen 2009). Netizens can also use HFS to manipulate other netizens' understanding of certain issues by providing false information or creating rumors. Due to the different motives of netizens, information provided might, initially, be selective or biased (Liu and Ling 2012; Hao 2009; Chen 2011). Although the whole picture might become clear over time when more information is provided, it could already have caused damage such as bullying and reputational harm to the targeted innocent person. The online trial then becomes a trial without rationality or neutrality. It can become a personal tool to attack others.

HFS can sometimes be used to form norms or moral standards when undertaken by collective interest groups on specific issues. Examples of this include the middle-finger Hsiao case and the cat abuse case referred to above. As these types of cases attract the public and media's attention, it then attracts more comments and judgments from netizens in both the cyber world and from the media in the real world (Chen and Guo 2009). This can lead to penal populism with people targeted by HFS suffering from more severe penalties being imposed as the public places pressure on judges to make decisions based on the moral standards dictated by the netizens' outrage on issues. Furthermore, collateral damage can be caused such as in the case of Hsiao where university colleagues and family members were also harassed as information about them was also disclosed online. HFS can, accordingly, be uncontrollable and can

impact the criminal justice system and create conflict and social dishar-
mony (Li and Li 2009; Ye and Li 2009).

Another important consequence of HFS relates to invasion of privacy.
Privacy involves the protection of any private activity that an individual
would not like to share with others, including personal messages, lifestyle, and income (Tang 2010; Gao 2012). While most information provided through HFS is information that can be found from any online
search engine, other information may be private to an individual netizen who knows the target or lives nearby. Although privacy invasion is
sometimes justifiable if a case relates to the public good and prevention
of crime, it may still cause irreversible harm to the target and his or her
family. Better rules are required to balance the invasion of privacy and
the public interest.

Cyber crowdsourcing as online vigilantism

"Cyber crowdsourcing" as used here is sometimes referred to by scholars
as "cyber vigilantism" or "online vigilante justice" (Ong 2012). Vigilantism refers to ordinary people attempting to rectify the structural flaws
in society to achieve social justice (Brown 1975) or, in colloquial terms,
ordinary people taking the law into their own hands. It is a reaction
to deviance seen as sufficiently severe to warrant the attention of the
formal justice system that is not being adequately addressed by official
agencies. Flaws and injustice in the formal justice system are present in
every society. Vigilantes act as informal community guardians to offset
the inadequacies of that system (Burrows 1976). Vigilantes believe that
the police and other authorities will not, or are unable to, handle certain cases and thus seek to redress such cases on their own to punish
rule-breakers who have brought shame to the entire society. Chang and
Poon (forthcoming) argue that a netizen who participates in online vigilantism 1) perceives the formal justice system to be ineffective and thus
seeks to achieve social justice by remedying the flaws in that system; 2)
believes that he or she has the ability to make society better by acting as
an informal guardian of it; and 3) uses the Internet and social networking platforms as a novel means of achieving social justice and punishing
deviants who slip through the formal justice system.

Public-private cooperation in criminal investigations is growing, with
some cybercrime investigations being conducted with the support of
financial institutions and information security companies (Chang
2012a, 2012b). Criminal investigations that rely on cyber crowdsourcing can be seen as "wikified" investigations. Tapscott and Williams

(2007) have argued that the "wiki" concept should not be limited to the software that enables people to edit a website such as Wikipedia. Rather, it should be used as a metaphor for the new era of collaboration and participation enabled by the Internet. Wiki use is based on four principles: being open, peering, sharing, and acting globally (Tapscott and Williams 2007). The concept has recently been adopted by academics, particularly those in the criminology arena, to describe cybercrime and cybercrime prevention (Chang 2012a; Wall 2008/2010) and to assess how the latter can be facilitated through mass collaboration and participation. Cyber crowdsourcing satisfies the four principles of wiki use and operates in a way similar to Wikipedia and Baidu Knowledge (Xu and Ji 2008). Cyber crowdsourcing platforms are usually open to all (although users sometimes need to register as a member), share the common goal of wanting to publicize certain issues, and rely on the willingness of netizens to contribute what they know. They are also global – anyone who is interested in the topic in question can contribute without geographical constraints, although there may of course be language constraints.

Such platforms are capable of achieving real results. For example, cyber crowdsourcers helped to identify the alleged perpetrators of the Boston Marathon bombings in 2013 (Glynn 2013), and they were also prominent in the sensational "Cat Abuse in Shun Tin Village" case in Hong Kong (Kuo 2012). Although cyber crowdsourcing may help investigators to collect information and evidence concerning crimes, there are many legal and ethical issues that need to be addressed before the approach can be adopted for use in formal criminal investigations. One concern relates to the identification of the wrong individual leading to considerable distress being caused to the individual and family members.

Conclusions

This chapter has examined cyber crowdsourcing as a popular online phenomenon in the Greater China region. The information from the archival study shows that HFS has been used to identify corrupt government officials, people who have engaged in minor offenses, and various forms of deviant behavior. It is sometimes used to satisfy curiosity about pop stars or other celebrities. Other cases have been used to publicize good deeds and acts of altruism that might otherwise have gone unnoticed. HFS is a double-edged sword that can, on one side, be used to expose and punish bad behavior and to maintain social justice, and, on the other, to damage people's lives and ruin good names and reputations of innocent people (Ruan 2011; Zhang 2008; Gao 2012).

It is apparent that HFS is not just a meaningless random activity but can help to solve crimes through public shaming and assisting police. In this sense, it can help society achieve social justice. Cases have already demonstrated the use of evidence from cyber crowdsourcing in criminal investigations, which demonstrates that private-public cooperation in criminal investigations is now becoming widespread.

Nonetheless, the negative impact that HFS might have on society should not be overlooked. Some of the potential negative effects of HFS include bullying, invasion of privacy, penal populism, reputational damage, and false identification of alleged criminals. In order to ensure that evidence collected via cyber crowdsourcing does not cause more harm than good to society, further regulation is needed to manage this emerging social phenomenon.

Bibliography

AppleDaily (2012) "Cat Abusing Case," *Apple Daily,* 11 November, http://hk.apple.nextmedia.com/news/art/20121118/18070785 (in Chinese) (accessed 8 May 2014).

Brown, R. (1975) *Strain of Violence,* New York: Oxford University Press.

Burrows, W. (1976) *Vigilante,* New York: Harcourt Brace Jovanovich.

Chang, L. Y. C. (2012a) *Cybercrime in the Greater China Region: Regulatory Responses and Crime Prevention across the Taiwan Strait,* Cheltenham: Edward Elgar.

Chang, L. Y. C. (2012b) "Responsive Regulation and the Reporting of Information Security Incidents," *Issues and Studies* 48(1): 85–119.

Chang, L. Y. C and Poon, R. (forthcoming) "Online Vigilantism: Attitudes and Experiences of University Students in Hong Kong," *International Journal on Offender Therapy and Comparative Criminology.*

Chen, H. F. and Guo, Y. J. (2009) "Human Flesh Searching in Online Community," *Journal of BUPT* 11(4): 5–9 (in Chinese).

Chen, M. (2009) "Human Flesh Searching and Citizen Internet Monitoring," *Views* 3: 38 (in Chinese).

Chen, S. F. (2011) "Discussion on Mechanisms of Monitoring Freedom of Speech Freedom and Human Flesh Searching," *Journalism Lover* 18: 42–4 (in Chinese).

Cheng, J. W. and Xue, H. Z. (2011) "Use and Regulation of Human Flesh Search in Investigation of Case," *Social Science* 7: 114–21 (in Chinese).

Cheung, S. Y. (2009) "China Internet Going Wild: Cyber-hunting Versus Privacy Protection," *Computer Law and Security Review* 25, 275–9.

China Internet Watch (2014) "Weibo Had 167M Monthly Active Users in Q3 2014," http://www.chinaInternetwatch.com/10735/weibo-q3-2014 (accessed 7 January 2015).

Fu, R. (2009) "Brief Guide on China Online BBS Community," *China Internet Watch,* http://www.chinaInternetwatch.com/369/china-bbs-forums/ (accessed 14 December 2014).

Gao, S. M. (2012) "The Moral Problem Created by Human Flesh Search and Its Tackling Method," *Marketing Management Review* 7: 194 (in Chinese).

Glynn, C. (2013) "Boston Marathon Bombing 'Crowdsourcing': How Netizens Are Using the Internet to Help Solve Crimes," CBSNews, http://www.cbsnews .com/news/boston-marathon-bombing-crowdsourcing-how-citizens-are-using -the-Internet-to-help-solve-crimes/ (accessed 10 October 2014).

Hao, L. Y. (2009) "A Study on Cyber Violence and Human Flesh Search," *Journal of Teaching and Researching Exploration*: 220 (in Chinese).

Hatton, C. (2014) "China's Internet Vigilantes and the 'Human Flesh Search Engine'," 28 January, BBC, http://www.bbc.com/news/magazine-25913472 (accessed 10 October 2014).

Herold, D. K. (2011) "Human Flesh Search Engines: Carnivalesque Riots as Components of a 'Chinese democracy'," in *Online Society in China: Creating, Celebrating, and Instrumentalising the Online Carnival*, D. K. Herold and P. Marlot (eds.), Oxford, UK: Routledge, 127–45.

Ju, Y. (2006) "The Loss of Privacy in New Media Era," *Xinhua News*, 28 June, http://news.xinhuanet.com/newmedia/2006-06/28/content_4762447_6.htm (in Chinese) (accessed 27 May 2014).

Kuo, S. (2012) "Hong Kong Juveniles Arrested for Cat Abusing," *China News*, http://www.chinanews.com/ga/2012/11-19/4339531.shtml (in Chinese) (accessed 20 September 2014).

Li, Y. and Li, D. X. (2009) "Productive Power of Moral Discourse and Rise of Chinese Type Cyber Manhunt," *Journal of Zhejiang University* 5: 1–9 (in Chinese).

Liu, B. U. and Ling, H. Y. (2012) "Discussing the Development of Social Incidents in the Media of Internet in the View of Sociological Perspectives," *Modern Communication* 9: 111–15 (in Chinese).

Liu, D. M. (2011) "Discuss the Characteristics and Effectiveness of Human Flesh Search" *Hei Long Jiang Shi Zhi* 3, 54–6 (in Chinese).

Liu, N. N. and Hao, H. K. (2009) "A Study on Applying Renrou Sousou to Investigative Practice," *Journal of Chinese People's Public Security University* 6: 9–16 (in Chinese).

Liu, W. F. (2009) "Research on Achieving Cyber Justice or Cyber Violence of Human Flesh Search," *Journal of Hubei University of Economics* 6(10): 130–2 (in Chinese).

Liu, Z. J. and Zhuang, M. (2012) "A Study of Internet Human Powered Search Engine," *Journal of Sanming University* 29(5): 82–7 (in Chinese).

Lu, Y. N. (2012) "A Study on Human Flesh Search," *Journal of Jiannan Literature* 5: 230 (in Chinese).

Ong, R. (2012) "Online Vigilante Justice Chinese Style and Privacy in China," *Information and Communications Technology Law* 21(2): 127–45.

People (2008) "Government Official Molestation Case," 4 November, http://politics.people.com.cn/GB/30178/8278124.html (accessed 8 May 2014).

Reuters (2008) "Hong Kong's Edison Chen Quits After Sex Scandal," 21 February, http://www.reuters.com/article/2008/02/21/us-hongkong-photos -idUSHKG36060820080221 (assessed 26 February 2015).

Ruan, J. S. (2011) "Discuss the Essentiality and Legality of Human Flesh Search," *Journal of Guangxi Police Academy* 1: 55–7 (in Chinese).

Shi, W. C. and Chi, W. M. (2009) "A Study on Juvenile's Human Flesh Searching Activities," *Theory Horizon* 3: 178–9.

Shu, Y. Y. (2012) "Discuss the Pros and Cons of Human Flesh Search," *Jian Nan Literature* 1: 169 (in Chinese).

Steiner, P. (1993) "On the Internet, Nobody Knows You're a Dog," *New Yorker*, 5 July, http://en.wikipedia.org/wiki/File:Internet_dog.jpg (accessed 25 February 2015).

Taipei Times (2010) "University Student Under Fire for Ambulance Incident," 30 December, http://www.taipeitimes.com/News/taiwan/archives/2010/12/30/2003492244 (accessed 26 February 2015).

Tang, X. T. (2010) "Criminologists Should Study on How to Regulate Crowdsourcing," *Journal of Hunan Public Security Academy* 4: 30–4 (in Chinese).

Tapscott, D. and Williams, A. D. (2007) *Wikinomics: How Mass Collaboration Changes Everything*, London: Atlantic Books.

Wadhwa, T. (2013) "Lessons from Crowdsourcing: The Boston Bombing Investigation," *Forbes*, http://www.forbes.com/sites/tarunwadhwa/2013/04/22/lessons-from-crowdsourcing-the-boston-marathon-bombings-investigation/ (accessed 12 November 2014).

Wall, D. S. (2008/2010) "Cybercrime and the Culture of Fear: Social Science Fiction(s) and the Production of Knowledge About Cybercrime," revised May 2010, *Information, Communication and Society* 11(6): 861–84.

Wang, F. Y., Zeng, D., Hendler, J. A., Zhang, Q. P., Feng, Z., Gao, Y. Q., Wang, H. and Lai, G. P. (2010) "A Study of the Human Flesh Search Engine: Crowd-powered Expansion of Online Knowledge," *IEEE Computer Society* 43(8): 45–53.

Wang, L. Y. and Liu, R. (2009) "The Broadcasting Process and Its Problem of Human Flesh Searching," *Today's Mass Media* 2: 73–4 (in Chinese).

Wang, S. Y. (2009) "Human Flesh Search Engines: Production and Self-discipline of Organization," *Online Communication Research* 8: 91–5.

Xiao, P. (2011) "An Analysis of the Phenomenon of Internet Mass Hunting," *Journal of Hunan University* 25(1): 156–60.

Xu, B. and Ji, S. (2008) "Human Flesh Search Engine: An Internet Lynching?," *Xinhua News*, http://news.xinhuanet.com/english/2008-07/04/ (accessed 27 June 2014).

Xu, H. W. and Li, S. F. (2011) "Discuss the Reason Why Human Flesh Search End Up with Cyber Violence," *Science and Technology Information* 22: 30–1 (in Chinese).

Yang, G. and Zhang, Z. C. (2010) "Analysis of the Privacy Issues of Human Flesh Search Engine," paper presented at the 2010 International Conference on Computer and Communication Technologies in Agriculture Engineering, Chengdu, 12–13 June.

Ye, X. M. and Li, J. (2009) "Moral Analysis of Internet Human Flesh Search," *Youth Studies* 5: 43–6 (in Chinese).

Yuan, G. S. and Zhao, Z. G. (2012) "Analysis of Flesh Search from the Communication Perspective," *Journal of Hebei University of Technology* 4(2): 76–80.

Zhang, X. (2008) "Why Human Flesh Search Cannot Be Controlled," *China Society Periodical* 21: 57–8 (in Chinese).

17
Hacktivism and Whistleblowing in the Era of Forced Transparency?

Alana Maurushat

Leni Riefenstahl

Leni Riefenstahl is one of the most controversial women and artists of the last century. Painter, actress, dancer, director genius, personal friend of Hitler, and commissioned film director of the Propaganda Ministry in Nazi Germany, Riefenstahl is both reviled and revered. She is best known for her direction of two films, *Triumph of the Will* (Riefenstahl 1935) and *Olympia* (Riefenstahl 1938).

Hitler commissioned Riefenstahl to make a feature-length documentary of the Nuremberg Nazi Party Rally of 1934, which became known as *Triumph of the Will*. The film won the National Film Prize and a gold medal at the Venice Biennale. It is considered to be one of the most significant and influential propaganda films ever made. In the opening shot, the Fuhrer descends from the clouds, leading into sequences resembling a travel documentary of Nuremberg set to the music of Wagner. She cuts from long shots to close-ups throughout the film, with the Fuhrer centered in many scenes, allowing the viewer the occasional cinematic illusion of seeing the sights through Hitler's eyes. The net effect is an idolization of Hitler as a magnanimous and charismatic leader.

Olympia fuses aesthetics with athletics in an artistic documentary of the 1936 Berlin Olympics. In the words of Riefenstahl herself:

I could see the ancient ruins of the classical Olympic sites slowly emerging from patches of fog and the Greek temples and sculptures drifting by: Achilles and Aphrodite, Medusa and Zeus, Apollo and Paris, then the discus thrower of Myron. I dreamed that this statue changed into a man of flesh and blood, gradually starting to swing the discus in slow motion. The sculptures turned into Greek temple dancers dissolving in flames, the Olympic fire igniting the torches to be carried from the Temple of Zeus to the modern Berlin of 1936 (Knaus and Riefenstahl 1992, p. 171).

Some critics have labeled *Olympia* as a Nazi propaganda film, citing examples of Riefenstahl's close-up shots of Hitler along with the triumphs of German athletes in the 1936 Olympics. Others have argued against this classification, drawing on Riefenstahl's portrayal of non-Aryan triumphs in the games, most famously a close-up of Hitler's facial expression as African American Jesse Owen crosses the finish line to win the gold medal. Unquestionable is Riefenstahl's contribution to innovative filmmaking techniques which have been heralded along the lines of Hitchcock and Welles. Unusual camera angles, extreme close-ups of facial expressions, and the use of a camera on a railway track to film both the athletes in motion alongside the reaction from the crowd, were ground-breaking techniques at the time and have since become industry standards.

Film critics around the world were unanimous in their praises for the cinematic aspects of *Triumph of the Will* and *Olympia*. In her autobiography, Riefenstahl reprints many of the media's reviews of *Olympia*, perhaps excessively so. This narcissistic tone continues in her book with elaborate recounts of the thundering of applause as she went from one European city to another for the debut of her films. She writes:

The end of the film was greeted by long, indeed almost endless applause... When Hitler thanked me and handed me a lilac bouquet I felt faint – and then lost consciousness. Premiere of *Triumph of the Will* (Knaus and Riefenstahl 1992, p. 166).

There was already some clapping during the prologue, and it was repeated over and over. Berlin premiere of *Olympia* (Knaus and Riefenstahl 1992, p. 224).

Never in my life, before or after that, have I received so many and such wonderful bouquets of flowers. Vienna screening of *Olympia* (Knaus and Riefenstahl 1992, p. 225).

...the enthusiasm was even more wild ... Hundreds of young girls in Styrian costumes formed a guard of honour from our hotel to the cinema. Graz debut of *Olympia* (Knaus and Riefenstahl 1992, pp. 225-6).

Riefenstahl's American tour of *Olympia* was markedly different from the showers of flowers and flowery praise encountered on her European tour. While film critics continued to praise the artistic merit of *Olympia*, the newspapers were heavily focused on the dreadful events of Kristallnacht (Night of Broken Glass). On 9 November 1938, Nazi troopers, SS, and Hitler Youth members were incited by Hitler and Propaganda Minister Joseph Goebbels to "rise in bloody vengeance against the Jews." Kristallnacht was a massive, coordinated attack across the German Reich where synagogues were burned, Jewish homes and businesses were broken into and vandalized, while many Jewish people suffered brutalization at the hands of their attackers. It is estimated that 25,000 Jewish men were rounded up and shortly afterwards sent to concentration camps (The History Place 1997). The world was shocked and outraged by these events, creating a storm of negative publicity for Hitler's Germany. The United States recalled its ambassador to Germany shortly after these events. Riefenstahl felt the reverberations of public outcry. In Hollywood, many studio directors and producers would not meet publicly with her, and the Anti-Nazi League called for a boycott against her. Signs were plastered on Hollywood's streets reading, "there is no place in Hollywood for Leni Riefenstahl" (Knaus and Riefenstahl 1997, p. 239). Still, Riefenstahl continued to be revered by film spectators and critics. That same year, *Olympia* won the prestigious 1938 Venice Film Festival.

Post-war treatment of Riefenstahl was a mixture of scorn and admiration. During the final months of the war, when American and Russian troops descended upon Berlin, she headed to what she thought would be safer ground in the countryside. She recounts doors slamming in her face and the cold reception she received by many who knew her, including her former lover and various past employees and friends who accused her of being a Nazi: "You Nazi Slut!" to be precise (Knaus and Riefenstahl 1992, p. 302). This type of tension is continually seen in Riefenstahl's treatment following the war, where she was imprisoned on a number of occasions by American and French guards, including a sojourn at the so-called "Bear Cellar," where major military figures were held. In spite of her association with the Nazi Party and the interrogations endured, Riefenstahl continued to be revered, as exemplified by prison guards telling her the locations of unguarded holes in the prison's fences in order to ingratiate themselves with the great and famous director. Sent to a mental institution, left in poverty with her property confiscated, Riefenstahl was placed under arrest and investigated by post-war authorities. She was found to be a Nazi sympathizer but not an activist. As such, she was never convicted of any crime, neither for her involvement with Nazi propaganda nor for the use of gypsy inmates from a concentration camp for the making of her film *Tiefland.*

Riefenstahl continues to be reviled by many for her personal relationship with Hitler and for complicity with the Propagation Ministry. Hitler's rise to supremacy has often been attributed to his successful propaganda campaign. It remains to be seen whether Riefenstahl's association with Hitler and the actions of the Nazi Party were feigned indifference, mere ignorance, or wilful blindness. She consistently claimed that she was completely unaware of the situation in the concentration camps and Hitler's involvement with the extermination of those deemed unworthy of human life. Arguably, she remained ignorant for quite some time. This is best summed up with an extract from Riefenstahl's memoirs reflecting on her acquired knowledge of Hitler's role in the concentration camps:

> I was totally confused. Perhaps, I thought, Hitler changed so much because of the war, perhaps because of the isolation in which he lived from the beginning of the war. From that moment on he was out of touch with the people beyond his own entourage. During his earlier rallies he had been infected by the emotions of the cheering crowds, absorbing them as a medium receives messages; positive impulses were transferred to him, suppressing the negative in him. After all, he wanted to be honoured and loved, but in his voluntary isolation there were no more human relationships. ... That was my attempt at finding an explanation for his schizophrenic nature (Knaus and Riefenstahl 1992, p. 314).

Riefenstahl's obsession as an artist to create something perfect and transcendent through time allowed her to fall under Hitler's spell in what could arguably be seen as wilful blindness, if not altogether complicity with the cataclysmic force of the Nazi Party. She writes, "suddenly I thought of Hitler's birthday [for the world premiere of *Olympia*] and said impulsively, 'Wouldn't 20 April be a good date.' ... 'You know, said Hitler, 'we'll schedule the premiere for 20 April after all, and I will come, I promise you.'' Incredulous, I stood there, unable to utter a single word" (Knaus and Riefenstahl 1992, p. 222). Indeed, as a director with an acute eye for detail, one has to wonder that she did not question the state of appearance of her concentration camp workers in the film *Tiefland.* Riefenstahl will forever remain revered and reviled.

Forcing transparency: Will history revere or revile hacktivists?

Civil activists in the 1960s and 1970s had sit-ins and protests for civil rights and against war. Many people thought that this civil disobedience could lead to change and open governments. This would lead to rational and critical discussion of citizens with governments in a move move toward a more open and transparent democratic governance. In the late 1970s and early 1980s, many governments enacted laws around freedom and access to information to ensure better government transparency. Canada enacted freedom of information laws in 1983 (Holsen 2007) while Australia's freedom of information act entered into force in 1982 (Office of the Australian Information Commissioner 2014). Prior to such enactment of freedom and access to information, it was difficult to obtain copies of government documents. These laws were devised in an attempt to move the disclosure of information default from private to public. In this sense, a government employee would not ask when something should be made public, but rather, when something should be made private. In other words, transparency by default.

While freedom and access to information laws have shifted the line of transparency, they did not achieve transparency by default. Internal guidelines for when information should remain private or public were muddled with bureaucratic wording and confusion. The end result was that government employees began to self-censor. This took place in two main ways. The first, when classifying documents, employees erred on the side of caution and thus over-classified documents as private/secret and under-classified documents as public/transparent. The second, when access to information requests were made to officials, often documents were so blacked out that it was difficult to ascertain with any certainty what decision or policy was adopted or why. The permanent black pen effect began. Transparency was not achieved.

The first part of the twenty-first century will go down in history as the era when hacktivists opened governments. The line of transparency is moving by force. The Twitter page for WikiLeaks demonstrates this ethos through its motto "we open governments" and its location:"everywhere." Hacktivism is a form of civil rights activism in the digital age. In principle, hacktivists believe in two general but spirited principles: respect of human rights and fundamental freedoms including freedom of expression and personal privacy, and the responsibility of government to be open, transparent, and fully accountable to the public. In practice, however, hacktivists are as diverse in their backgrounds as they are in their agendas.

In its traditional form, hackers were people who used clever technical solutions to solve problems. Hacktivism, then, is the use of a clever technical solution in support of a cause. Hacktivism is also referred to as electronic civil disobedience, ethical hacking, and politically motivated computer crime.

Hacktivism is not new. In the late 1980s, Australian hacktivists penetrated the NASA computer system, releasing a worm known as WANK – Worms Against Nuclear Killers. The worm was written and released as a form of protest for NASA's launch of the rocket Galileo which was to navigate itself to Jupiter using nuclear energy (Dreyfus and Assange 2011). The infamous German hacker group Chaos Club was also busy in the late 1980s attacking German government systems to protest against the collecting and storing of census information; the groups believed that the government should not collect or store the personal information of its citizens (Dreyfus and Assange 2011).

Moving forward to the first decade of the twenty-first century, hacktivism, while not new, has fundamentally changed in one distinct manner – the ability to participate in attacks (denial of service attacks) is no longer limited to an elite group of people with excellent computer skills; the technology is available to the masses in an accessible format for those with limited technical skill. Click the download button for LOIC software. Choose the demonstration you wish to participate in, then click again. *Fait accompli.*

People around the globe are participating in denial of service attacks of a variety of websites for a variety of causes. Websites that have been attacked to date include the Australian Parliament's website, PayPal, Mastercard, Mexican government website, paedophilia websites, the New York Stock Exchange, the Toronto Stock Exchange, News of the World, Oakland City Police, Ecuadorian government, Peruvian government, and the list goes on and on.

The most well-known hacktivist organization is Anonymous. Anonymous is the umbrella group for a plethora of small groups that operate under it. In addition to performing denial of service attacks, members of some of the smaller groups participate in more sophisticated forms of hacktivism that require a higher range of computer skills. Instances of these more sophisticated attacks include the release of names and details of the Mexican drug cartel, Los Zetas, the names and details of individuals who use child pornography sites, and the capturing of secret documents held by governments around the world – some of these documents are then given to and released by WikiLeaks (Maurushat forthcoming).

Hacktivism isn't limited to attacking information systems and retrieving documents. It also extends to finding technical solutions to mobilize people. At the height of the Egyptian e-revolution, the major Internet service providers and mobile phone companies shut down the Internet (the kill switch) preventing people from using the Internet and mobile phones (Maurushat, Chawki, Shazly and Al-Alosi 2014). This, in turn, affected the people's ability to mobilize. Anonymous worked around the clock to ensure that images from the revolution were still being sent to the international press. Hacktivists also work to penetrate the Iranian government's firewall to tunnel passages through Iranian Internet access, allowing Iranian citizens to view blocked sites (Maurushat forthcoming). There are similar initiatives for China and Saudi Arabia. Keeping secrets and preventing citizens from accessing information is no longer an achievable goal. The question becomes, should governments adopt heavy-handed policies and law to investigate and prosecute hacktivists to deter such activity and keep the status quo? Or should governments enact an appropriate legislative response that reflects the reality of this new era – the forced line of transparency?

We are at a point in history when citizens around the world are insisting that governments operate in a more open and transparent fashion. This has meant a growing trend of the use of open-source software, open access to journals and research, and open access and accountability of governments for their actions. This insistence on transparency and accountability has been seen time and time again in the leaking of war documents from the Middle East to WikiLeaks; the Snowden revelations of American surveillance by the NSA; the Arab Spring uprisings in Tunisia, Egypt, Yemen, and Syria; the Occupy movements around the world; and the various acts and exposures of ethical hacking such as some of the online protests and hacktivism efforts of Anonymous. Some of the above incidents have involved whistleblowing and/or hacktivism.

There is a growing movement of ethical hacking to retrieve confidential documents which are later posted to online leak sites such as WikiLeaks. Meanwhile online websites such as the Social Hack Leak Directory Project (2014) provide a wide range of information on leak sites, how to leak, and how to use anonymizing technology, and provide evolving leak technologies and information on how to ethically hack (Social Hacking Leak Directory Project 2014). In Italy, a group of computer scientists developed MafiaLeaks in a similar vein to remove the secrecy about the mafia; the site guarantees anonymity (Francheschi-Bicchierai 2013).

Whistleblowing is the disclosure of illegal, immoral, or illegitimate practices of an organization by a member or employee of the organization (Brown, Gobert and Punch 2000). Disclosure could be to the media, to a regulatory authority, or to the public in general (such as via disclosure on a website). Whistleblowing involves the disclosure of otherwise confidential information where it is a matter of the "public interest." Many jurisdictions have enacted legislation that shields a whistleblowing member or employee of a government, corporation, or organization from criminal sanction and legal liability. As will be seen, this protection isn't, however, absolute.

The concepts of external and internal whistleblowing are somewhat confusing (Morehead, Dworkin and Baucas 1998). The terms "external" and "internal" refer to the recipients of the information and not to the person who exposes the information. An internal whistleblower is a member or employee of an organization who sends leaked information to someone within the organization. External whistleblowing occurs when the person chooses to share the information with someone external of the organization. In some jurisdictions, both internal and external whistleblowers are protected under the law, while in other jurisdictions the recipient must be internal.

Whistleblowers enjoy legal protection in many jurisdictions, including Australia. Depending on the jurisdiction in question, whistleblowers are protected from criminal charges, civil liability, and being fired for disclosing information about corrupt, illegal, or immoral practices of federal governments, state governments, and corporations.

When someone external to an organization exposes wrongdoing he or she is not considered a whistleblower and is not shielded from criminal sanction and legal liability.

Third parties, therefore, are not protected by whistleblower legislation. If a hacktivist, for example, obtains a document by gaining unauthorized access to a computer, they are not considered a whistleblower under legislation. The primary goal of whistleblowing legislation is to reduce – if not prevent – retaliation for exposure of malpractice or wrongdoing in the workplace. The goal of whistleblowing has never been given a broad interpretation to cover third parties. Regardless of who and why someone blows the whistle, whether they be a government employee or ethical hacker, the goal remains essentially the same – to expose a wrongdoing.

In a typical scenario, an ethical hacker/whistleblower will access without authorization a database to retrieve information on corrupt practices.

This information will then be published to a website, given to a newspaper, and/or submitted to a leak site. This unauthorized access of data, a database, or computer will constitute a criminal offense in Australia. Most jurisdictions have enacted computer-related offenses which are often referred to as unauthorized access, modification, or interference to data systems or electronic communications. Such criminal provisions generally address situations where any component of a computer (hard drive, software, network) is tampered with, allowing for unauthorized access, modification, impairment, or interference to data or a data system. The very nature of hacking – whether it be to expose corrupt practices or out of mere curiosity – involves the exploration (and sometimes exploitation) of vulnerabilities which, at a minimum, involve unauthorized access to data. There are no security research, whistleblowing, or public interest exemptions to criminal computer offenses in any jurisdiction (Maurushat forthcoming).

One thing is for certain, how civil rights will apply to hacktivism and third-party whistleblowing is anything but certain. There are no exceptions to the cybercrime/computer crime provisions in most jurisdictions around the globe. How will the courts maneuver in this new era of activism meets computer? The recent judgment of a German court gives us some indication of how Western democratic courts may view hacktivism. In 2001, two German civil rights activist groups, Libertad and "Kein Minch ist illegal" had called for protests against Lufthansa for their policy of helping to identify and deport asylum seekers. There was an offline protest at the Lufthansa shareholders meeting. This was met with an online protest. The online protest consisted of a DDoS attack where over 13,000 people participated, shutting down Lufthansa's server for two hours (this is pre LOIC) (libertad.de 2006).

One of the protest organizers, Andreas-Thomas Vogel, was convicted of coercion by a regional German court. On appeal, the higher court found that there was no coercion under §240 of the German criminal law. They reasoned that there was no violence or threatening behavior. Further, the court reasoned there needs to be a permanent and substantial modification of data to be found guilty of incitement of alteration of data. The court viewed the DDoS attack as a modern form of non-violent blockade fully within the right to freedom of expression.

In Australia, a similar attack attracted comments from the court as falling within terrorist activity. There was no mention of freedom of expression or freedom of assembly or a right to protest. Matthew George was an Australian member of Anonymous who participated in "Operation Titstorm" as a form of protest for government censorship. He was charged and convicted of incitement. The magistrate stated that George had incited others to attack government websites and went so far as to liken his activities to cyber terrorism – a claim that is truly outrageous given the context of the protest. George was given a $550 fine. George was not a ringleader but merely a participant using LOIC software. Taken from statements made by George to the *Sydney Morning Herald*, he stated:

> We hoped to achieve a bit of media attention to why internet censorship was wrong ...
>
> I didn't think that I would ever get caught. I was actually downloading connections from other computers in America, so I didn't think the Australian government would be able to track me down.
>
> I had no idea that what I was doing was illegal. I had no idea that there was incitement and it was illegal to instruct others to commit a legal [sic] act (Whyte 2011).

There is an underlying theme here whereby many DDoS users do not realize that they are participating in an illegal activity. In Australia, the court took a decidedly different view to online civil disobedience. Newcastle resident 22-year-old Matthew George received a $550 fine for inciting others to perform an illegal act when he participated in the defacement of the Australian Parliamentary website as a form of protest against newly proposed Internet censorship law. Australians are not afforded the same safeguards and protections of civil rights as other Western democratic countries that have a strong bill of rights or nationally protected human rights enshrined in a constitution.

It remains to be seen if future courts will look at hacktivism and third-party whistleblowing in a different light. Table 17.1 offers a comparison between online and offline activism.

Table 17.1   Comparison between offline and online activism

Offline	Online
Sit-ins	Virtual Sit-ins
Barricades	Denial of Service Attacks & Website Redirection
Political Graffiti	Website Defacements
Wildcat Strikes	Denial of Service Attacks & Website Redirection
Underground Presses	Site Parodies, Blogs, Facebook and Twitter Protests
Petitions	Web Petitions, Facebook Likes

While the courts have yet to debate the parallels between online and offline activism, there has recently been a number of conflicting trends in the Australian landscape toward transparency. The first trend relates to information disclosure protection legislation. On 15 January 2014, the *Public Interest Disclosure Act* 2013 (Cth) entered into force, providing protection to whistleblowers. Previously, whistleblowing protection was only available to those disclosing information from state government initiatives; the protection now extends to the Commonwealth government. The protection does not, however, extend to private entities. Take, for example, the lesson that Freya Newman learned after disclosing a rare and possible unmerited $60,000 scholarship given to Prime Minister Tony Abbott's daughter at a design school whose chairman, Les Taylor, is purported to be a Liberal Party donor. Though the leak of the scholarship has direct bearing in that there is public interest in knowing the relationship between the design institute, the chairman, the prime minister, and the Liberal Party, Newman will not be afforded protection for disclosure, as Australia's whistleblowing legislation does not extend to private corporations (Safi 2014).

Following the leaks of United States National Security Agency documents by former contractor Edward Snowden in June 2013, it was reported that Australia law enforcement has received information from the NSA surveillance programs. It is further believed that the Attorney General's Department is seeking the power to "break into anonymisation and encryption software like Tor" (Keane 2013). One concern is the recently enacted *National Security Legislation Amendment Act (No 1)* 2014 (Cth), which threatens journalists and whistleblowers with a ten-year prison term if they publish classified information. These provisions have already entered into force and have been widely critiqued for their potential effects on a free press.

On a more positive note, there seems to be a trend in some countries toward opening up data sets in what could best be described as open data policies (Davies 2014). While open data has been defined differently for various purposes, there is a trend toward interpreting open data as data that would be available under a freedom of information request, after redacting all data that is subject to disclosure exemptions for reasons such as containing personal information, security issues, being cabinet material, or being subject to commercial confidentiality.

Should we keep secrets in times of weak law, and should the law do more to support secrecy? Probably not. This is not to say that there aren't legitimate reasons for governments and corporations to keep secrets but that there needs to be better guidelines as to when documents should be made public and when they should remain private. The information culture of governments needs to change. And if governments don't change, hacktivists will continue to force the line of transparency. This is how I perceive the problem. Will hacktivists be revered or reviled? For the moment, governments view hacktivists as cybersecurity threats with the media reporting on hacktivism with a mixture of sentiment which is reflected in the public's view of hacktivism. As governments move forward, and the courts have the opportunity to grapple with many of the issues that hacktivists bring to light, I believe that hacktivists will be increasingly revered. In the words of WikiLeaks commenter and journalist Jay Rosen, "I don't have the answer; I don't even know if I have framed the right problem. But the comment bar is open so please help me out" (Rosen 2010).

Bibliography

Brown, A. J. (2014) *Whistleblowing in the Australian Public Sector: Enhancing the Theory and Practice of Internal Witness Management in Public Sector Organisations,* Canberra, Australia: ANU e Press, http://epress.anu.edu.au/anzsog/whistle-blowing/mobile_devices/index.html (accessed 10 February 2014).

Davies, T. (2014) "Open Data Policies and Procedures: An International Comparison," http://dx.doi.org/10.2139/ssrn.2492520 (accessed 23 October 2014).

Dreyfus, S. and Assange, J. (2011) *Underground,* Sydney: Random House.

Francheschi-Bicchierai, L. (2013) "MafiaLeaks: Inside the Site That Wants to Take Down the Mob," http://mashable.com/2013/11/07/mafialeaks/ (accessed 2 October 2014).

Gobert, J. and Punch, M. (2000) "Whistleblowers, the Public Interest, and the Public Interest Disclosure Act 1998," *Modern Law* Review 63(1): 25–54.

Holsen, S. (2007) "Freedom of Information in the U. K., U. S. and Canada," *The Information Management Journal,* May/June: 50–5, http://www.arma.org/bookstore/files/Holsen.pdf (accessed 22 September 2014).

Keane, B. (2013) "The Greatest Threat to Our Rights Is the Attorney-General's Department," http://www.crikey.com.au/2013/06/05/the-greatest-threat-to-our-rights-is-the-attorney-generals-department/ (accessed 15 May 2014).

Knaus Verlag Gmbh, A. and Riefenstahl, L. (1992) *Leni Riefenstahl: A Memoir,* New York: Picador.

Libertad.de (2006) File Reference 1 Ss 319/05, 22 March 2006, Frankfurt am Main Higher Regional Court, http://www.libertad.de/service/downloads/pdf/olg220506.pdf (accessed 10 September 2014). The translation of the decision was provided courtesy of UNSW LLM student Bodil Diederichsen.

Maurushat, A. (2012) "Forced Transparency: Should We Keep Secrets in Times of Weak Law, and Should the Law Do More to Support Secrecy?," *Media & Arts Law Review* 17(2): 239–42.

Maurushat, A., Chawki, M., Al-Alosi, H., el Shazly, Y. (2014) "The Impact of Social Networks and Mobile Technologies on the Revolutions in the Arab World – A Study of Egypt and Tunisia," *Laws* 3(4).

Maurushat, A. (forthcoming) *Ethical Hacking,* Ottawa: University of Ottawa Press.

Morehead Dworkin, T. and Baucas, M. (1998) "Internal vs. External Whistleblowers: A Comparison of Whistleblowing Processes," *Journal of Business Ethics* 17, 1281–98.

Office of the Australian Information Commissioner (2014) *About Freedom of Information,* http://www.oaic.gov.au/freedom-of-information/about-freedom-of-information (accessed 10 September 2014).

Riefenstahl, L. (1935) *Triumph of the Will* (film), Germany: Leni Riefenstahl -Produktion.

Riefenstahl, L. (1938) *Olympia* (film), Germany: Leni Riefenstahl-Produktion.

Rosen, J. (2010) "Jay Rosen's Press Think: The Afghanistan War Logs Released by Wikileaks, the World's First Stateless News Organization." 26 July. http://pressthink.org/2010/07/the-afghanistan-war-logs-released-by-wikileaks-the-worlds-first-stateless-news-organization/ (accessed 10 March 2015).

Safi, M. (2014) "Freya Newman Sentence Deferred Over Frances Abbott Scholarship Leak," *The Guardian,* 23 October, http://www.theguardian.com/australia-news/2014/oct/23/freya-newman-sentence-deferred-frances-abbott-scholarship-leak (23 October 2014).

Social Hacking Leak Directory Project (2014) http://events.ccc.de/congress/2011/wiki/SocialHacking_LeakDirectory (accessed 23 October 2014).

The History Place (1997) "Holocaust Timeline," http://www.historyplace.com/worldwar2/holocaust/h-knacht.htm (accessed 10 September 2014).

Whyte, S. (2011) "Meet the Hacktivist Who Tried to Take Down the Government," *Sydney Morning Herald,* 14 March, http://www.smh.com.au/technology/security/meet-the-hacktivist-who-tried-to-take-down-the-government-20110314-1btkt.html (accessed 10 September 2014).

Index

GPSR Compliance
The European Union's (EU) General Product Safety Regulation (GPSR) is a set
of rules that requires consumer products to be safe and our obligations to
ensure this.

If you have any concerns about our products, you can contact us on

ProductSafety@springernature.com

In case Publisher is established outside the EU, the EU authorized
representative is:

Springer Nature Customer Service Center GmbH
Europaplatz 3
69115 Heidelberg, Germany

www.ingramcontent.com/pod-product-compliance
Lightning Source LLC
LaVergne TN
LVHW012141040326
832903LV00004B/240